Synthese Library

Studies in Epistemology, Logic, Methodology, and Philosophy of Science

Volume 515

The aim of *Synthese Library* is to provide a forum for the best current work in the methodology and philosophy of science and in epistemology, all broadly understood. A wide variety of different approaches have traditionally been represented in the Library, and every effort is made to maintain this variety, not for its own sake, but because we believe that there are many fruitful and illuminating approaches to the philosophy of science and related disciplines.

Special attention is paid to methodological studies which illustrate the interplay of empirical and philosophical viewpoints and to contributions to the formal (logical, set-theoretical, mathematical, information-theoretical, decision-theoretical, etc.) methodology of empirical sciences. Likewise, the applications of logical methods to epistemology as well as philosophically and methodologically relevant studies in logic are strongly encouraged. The emphasis on logic will be tempered by interest in the psychological, historical, and sociological aspects of science. In addition to monographs *Synthese Library* publishes thematically unified anthologies and edited volumes with a well-defined topical focus inside the aim and scope of the book series. The contributions in the volumes are expected to be focused and structurally organized in accordance with the central theme(s), and should be tied together by an extensive editorial introduction or set of introductions if the volume is divided into parts. An extensive bibliography and index are mandatory.

Manuel Gustavo Isaac • Steffen Koch •
Kevin Scharp
Editors

New Perspectives on Conceptual Engineering - Volume 3

Applied Conceptual Engineering

Springer

Editors
Manuel Gustavo Isaac
GESDA - Geneva Science and Diplomacy Anticipator
Geneva, Switzerland

Steffen Koch
Department of Philosophy
Bielefeld University
Bielefeld, Germany

Kevin Scharp
University of Illinois Urbana-Champaign
Champaign, IL, USA

ISSN 0166-6991 ISSN 2542-8292 (electronic)
Synthese Library
ISBN 978-3-031-98534-8 ISBN 978-3-031-98535-5 (eBook)
https://doi.org/10.1007/978-3-031-98535-5

This work was supported by Swiss National Science Foundation, Grant number: P400PG_183807

This Springer imprint is published by the registered company Springer Nature Switzerland AG
The registered company address is: Gewerbestrasse 11, 6330 Cham, Switzerland

This collection of essays draws on lectures presented in the Conceptual Engineering Webinar Series. The webinar series was launched by Manuel Gustavo Isaac and Kevin Scharp in March 2020 at the University of St Andrews before being subsequently co-hosted with the University of Zurich. In 2022, it became an independent initiative run by the Conceptual Engineering Network (https://www.conceptualengineering.xyz), whose coordinators include Steffen Koch. The Conceptual Engineering Network curates the Conceptual Engineering YouTube channel (https://www.youtube.com/@ConceptualEngineering) for online dissemination.

Contents

Introduction xi

1 **How Language Teaches and Misleads: "Coronavirus" and "Social Distancing" as Case Studies** 1
Ethan Landes

2 **Psychiatric Kinds and the DSM: Notes from a Conceptual Building Site** 21
Rachel Cooper

3 **Conceptual Engineering of Medical Concepts** 43
Elisabetta Lalumera

4 **Rethinking the Ocean: Conceptual Engineering for the Next 10,000 Years** 69
Roberto Casati, Sacha Bourgeois-Gironde, Emilie Flamme, Alberto Gallace, Quentin Hiernaux, Stefano Malatesta, Marcella Schmidt di Friedberg, Enrico Squarcina, Colomban de Vargas, and Eva Wanek

5 **Conceptual Engineering, Semantic Tolerance and Flexibility** 95
Genoveva Martí and Lorena Ramírez-Ludeña

6 **Labeling Unlabeled Identities** 111
Katherine Ritchie

7 **Conspiracy Theories Are Not Theories: Time To Rename Conspiracy Theories** 137
Kevin Reuter and Lucien Baumgartner

Contributors

Lucien Baumgartner University of Zurich, Zürich, Switzerland

Sacha Bourgeois-Gironde Institut Jean Nicod, UMR 8129, ENS-EHESS-CNRS, DEC, Université PSL, Paris, France
Université Paris 2 Panthéon-Assas - Centre for Law and Economics, Paris, France

Roberto Casati Institut Jean Nicod, UMR 8129, ENS-EHESS-CNRS, DEC, Université PSL, Paris, France

Rachel Cooper Lancaster University, Lancaster, UK

Emilie Flamme Institut Jean Nicod, UMR 8129, ENS-EHESS-CNRS, DEC, Université PSL, Paris, France

Marcella Schmidt di Friedberg Università Milano Bicocca, Milan, Italy

Alberto Gallace Università Milano Bicocca, Milan, Italy

Quentin Hiernaux Institut Jean Nicod, UMR 8129, ENS-EHESS-CNRS, DEC, Université PSL, Paris, France

Elisabetta Lalumera University of Bologna, Department for Life Quality Studies, Rimini, RN, Italy

Ethan Landes University of Kent, Canterbury, UK

Stefano Malatesta Università Milano Bicocca, Milan, Italy

Genoveva Martí ICREA (Institució Catalana de Recerca i Estudis Avançats), Barcelona, Spain
Universitat de Barcelona, Barcelona, Spain

Lorena Ramírez-Ludeña Universitat Pompeu Fabra, Barcelona, Spain

Kevin Reuter University of Zurich, Zürich, Switzerland

Katherine Ritchie University of California, Irvine, CA, USA

Enrico Squarcina Université Paris 2 Panthéon-Assas - Centre for Law and Economics, Paris, France

Colomban de Vargas Sorbonne Université, CNRS, Station Biologique de Roscoff, UMR7144, ECOMAP & Plankton Planet, Roscoff, France
Research Federation for the study of Global Ocean Systems Ecology and Evolution, FR2022/Tara GOSEE, Paris, France

Eva Wanek Institut Jean Nicod, UMR 8129, ENS-EHESS-CNRS, DEC, Université PSL, Paris, France

Introduction

Conceptual engineering represents a sweeping movement within analytic philosophy, dedicated to assessing and improving our representational devices. In less than a decade, it has surged to the forefront of contemporary philosophical inquiry. This remarkable ascent owes to a confluence of factors. External determinants, such as securing substantial grants for collaborative research endeavors leading to seminal publications and fostering new synergies, the inception of international networks, the advent of online lecture series, and the creation of YouTube channels, have undoubtedly played a pivotal role in promoting it in the philosophy community. Equally significant are internal dynamics within conceptual engineering itself: By reconceptualizing philosophy as a normative examination of our representational apparatus, with ambitions both foundational and universal, it has been heralded as a "first philosophy" by proponents like Cappelen in the early days of the movement, which might have provoked many to engage with conceptual engineering, whether critically or not. Yet, perhaps the most conspicuous factor contributing to the movement's success is its very label.

The phrase "conceptual engineering" casts philosophy as a problem-solving endeavor geared towards real-world challenges, and promising tangible solutions. In so doing, it imbues philosophy with an applied, translational orientation at the core, countering stereotypes of impracticality and otherworldliness that have long plagued the discipline. Through conceptual engineering, philosophy comes to assume an outward-facing stance, turning its gaze toward practical problems beyond its conventional purview of addressing self-generated theoretical quandaries or other such intricacies.

This forms the topic of the third installment in the series *New Perspectives on Conceptual Engineering*. In contrast to the first two, this volume delves into applied issues stemming from the outside world rather than foundational ones (Volume 1) or those within specific subfields of philosophy (Volume 2). Titled 'Applied Conceptual Engineering,' it extends the scope of philosophical inquiry beyond its traditional boundaries.

In "How Language Teaches and Misleads: 'Coronavirus' and 'Social Distancing' as Case Studies," Ethan Landes explores an underexamined aspect of conceptual engineering: lexicalization. Utilizing the classical semiotic triangle (sign, meaning, reference), Landes introduces two critical properties for guiding lexical choices: decipherability, the ease with which users infer true information from a term, and misleadingness, the propensity for users to infer false information. Landes applies these concepts to a cost-benefit analysis of the terms "coronavirus" and "social distancing" during the COVID-19 pandemic, demonstrating the profound impact of lexical choices on public health safety when experts introduce new language.

Rachel Cooper, in "Psychiatric Kinds and the DSM: Notes from a Conceptual Building Site," adopts a literal engineering perspective to explore changes in "human kind' concepts crafted by expert committees. Challenging Cappelen's (2018) uncontrollability principle, Cooper highlights how DSM committees have successfully revised problematic concepts across editions of the Diagnostic and Statistical Manual of Mental Disorders. She addresses the unpredictability of the consequences of conceptual changes, emphasizing potential detrimental side effects despite good intentions, and contrasts the philosophical ideals of conceptual clarity, precision, and consistency with practical usage in scientific and practitioner communities. Cooper's insights from psychiatric classifications offer valuable lessons for conceptual engineers.

In "Conceptual Engineering of Medical Concepts," Elisabetta Lalumera examines two case studies—cancer and obesity—to illustrate the intricacies of revising conceptual classifications and demarcation criteria in medicine. She argues for the necessity of explicit conceptual engineering in medicine, where biological, clinical, and social considerations intersect and sometimes conflict. Lalumera emphasizes that medical concepts are inherently value-laden and should reflect consensus on practical and ethical choices, not just evidence. She details how conceptual engineering in medicine targets experts' conceptions—instead of semantically or psychologically construed concepts. Altogether, Lalumera concludes, this makes the medical domain particularly suitable for future projects.

Roberto Casati and colleagues, in "Rethinking the Ocean: Conceptual Engineering for the Next 10,000 Years," apply conceptual engineering to our understanding of the ocean. They argue that correcting misrepresentations of the ocean is vital for effective public policies addressing ecological crises. The authors analyze the network of concepts and metaphors shaping our view of the ocean, proposing alternative representations to mitigate harmful effects. They emphasize the need for a specific investigation into the viability, scientific adequacy, cognitive acceptability, and implementability of these reconceptualizations, underscoring the importance of making the ocean more visible and accurately represented.

Genoveva Martí and Lorena Ramírez-Ludeña's "Conceptual Engineering, Semantic Tolerance and Flexibility" brings metasemantic reflections to jurisprudence, extracting general lessons for conceptual engineering. They distinguish between semantic tolerance (coexistence of different uses for the same term) and semantic flexibility (expansion or contraction of a term's application). Through legal interpretation case studies, they conclude with a few caveats that, in the context of jurisprudence, semantic tolerance does not constitute conceptual engineering, whereas semantic flexibility does. They caution against the potential pitfalls of extensive conceptual engineering.

In "Labeling Unlabeled Identities," Katherine Ritchie examines the implications of labeling social identities in social, political, and academic discourse. She contends that identities exist independently of labels, though labeling can enhance their salience. Ritchie explores the effects of both labeling and not labeling identities, emphasizing the importance of the social dynamics between labeled and labeling groups. She concludes that labeling dominant and marginalized identities generally yields significant benefits and proposes strategies to mitigate negative impacts, challenging the notion in social policy that identity labels are inherently harmful.

Finally, in their chapter titled "Conspiracy Theories Are Not Theories: Time to Rename Conspiracy Theories," Kevin Reuter and Lucien Baumgartner challenge the epistemic validity of conspiracy theories by arguing that they do not meet the criteria typically associated with scientific theories—such as testing, confirmation, and refinement. Through corpus studies, they demonstrate that conspiracy theories function more like rumors and falsehoods than genuine theories. The authors advocate for relabeling these as 'conspiracy stories' or 'conspiracy narratives' to better reflect their nature and to avoid the misleading implication that they possess scientific legitimacy.

As this collection hopes to show, in a fast-transforming world, the need for conceptual engineering is more pressing than ever. Technoscientific advances, sociocultural innovations, and environmental changes reshape the fabric of our reality, disrupting fundamental concepts we use to navigate and act on it. Whether implementable or not, conceptual engineering here offers a means to regain control over how we make sense of things through representational devices. Ideally, it will evolve into a diverse, inclusive, and participatory endeavor toward better futures for all.

University of Illinois Urbana-Champaign — Kevin Scharp
Urbana-Champaign, IL, USA
Geneva Science and Diplomacy Anticipator — Manuel Gustavo Isaac
Geneva, Switzerland
Department of Philosophy, Bielefeld University — Steffen Koch
Bielefeld, Germany

Chapter 1
How Language Teaches and Misleads: "Coronavirus" and "Social Distancing" as Case Studies

Ethan Landes

Abstract The beginning of the COVID-19 pandemic offers a unique case study for understanding conceptual and linguistic propagation. In early 2020, scientists, politicians, journalists, and other public figures had to, with great urgency, propagate several public health-related concepts and terms to every person they could. This chapter examines the propagation of *coronavirus* and *social distancing* and develops a framework for understanding how the language used to express a notion can help or hinder propagation. I argue that anyone designing words, meanings, or concepts for propagation needs to appreciate the three-way causal relationship between language, people's mental representations, and the extramental world. Using this framework, I explore what makes "social distancing" a bad name for social distancing and why it is unproblematic that "coronavirus" is a scientifically loose way of speaking about the virus. Through this and further study of the many historical examples of linguistic and conceptual propagation, conceptual engineers can better understand the complex challenges facing people who want to propagate words, meanings, or concepts.

1.1 Introduction[1]

If conceptual engineers want to learn what strategies to employ—and pitfalls to avoid—while propagating words, meanings, or concepts, they need to look no further than science communication. The spread of scientific notions to everyday people provides a rich corpus of successful and not-so-successful efforts by

[1] Thanks to Deryn Thomas, Josh Habgood-Coote, Derek Ball, Manuel Gustavo Isaac, Kevin Scharp, Steffen Koch, the audience of the Conceptual Engineering Online Seminar, and anonymous referees for their feedback on this paper.

E. Landes (✉)
University of Kent, Canterbury, UK
e-mail: E.Landes@kent.ac.uk

M. G. Isaac et al. (eds.), *New Perspectives on Conceptual Engineering - Volume 3*, Synthese Library 515, https://doi.org/10.1007/978-3-031-98535-5_1

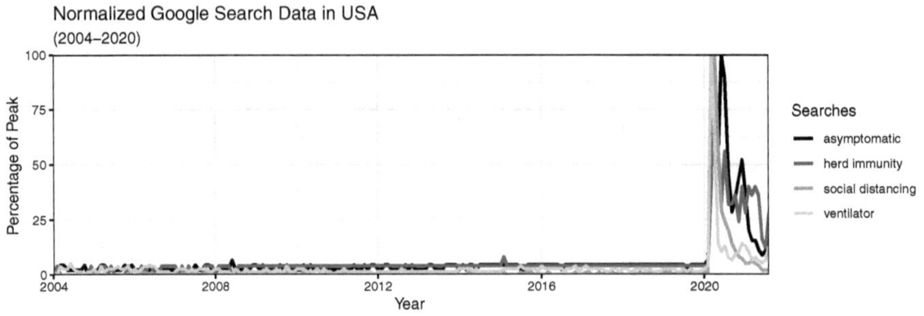

Fig. 1.1 Google search data in USA showing normalized peaks of terms related to COVID-19 from January 2004–December 2020. Data shows a large peak in early 2020. Each search term's percentage is normalized according to its peak popularity. Data sourced from https://trends.google.com/trends/. Graph by author in R using ggplot2 (Wickham, 2016)

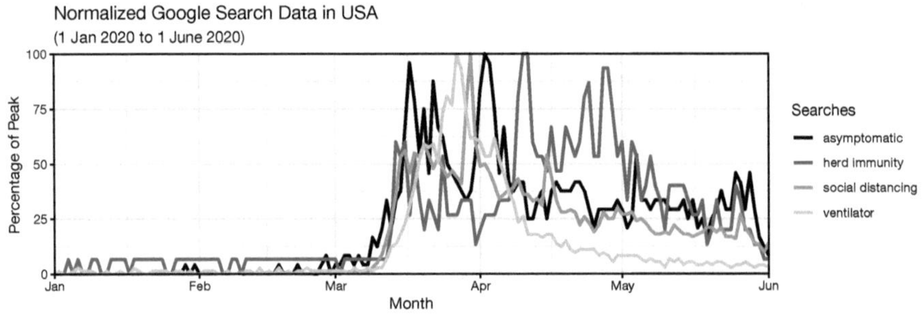

Fig. 1.2 Google search data in USA showing normalized peaks of terms related to COVID-19 from 1 January–1 June 2020. This figure shows the peak in Fig. 1.1 in more detail. Each search term's percentage is normalized according to its peak popularity. Data sourced from https://trends.google.com/trends/. Graph by author in R using ggplot2 (Wickham, 2016)

scientist, journalists, politicians, and other science communicators to change the vocabulary and conceptual repertoire of everyday people. To this end of learning more about conceptual and linguistic propagation from real case studies, this chapter looks at examples of propagation during the earliest stages of the COVID-19 pandemic. Over the course of just a few weeks in early spring 2020, the public was introduced to many pandemic-related terms, including "SARS-CoV-2", "social distancing", "COVID-19", "N95 masks", "herd immunity", "PPE", "ventilator", "respirator", "intubation", "asymptomatic", and "flattening the curve". These terms and the corresponding concepts, previously limited primarily to specialized scientific and medical settings, became part of people's everyday language (Figs. 1.1, 1.2 and 1.3). Compared to other cases of propagation, however, the early days of the COVID-19 outbreak were unique in the extraordinary speed and urgency by which technical language entered everyday speech over the span of mere days in March 2020 (Figs. 1.2 and 1.3).

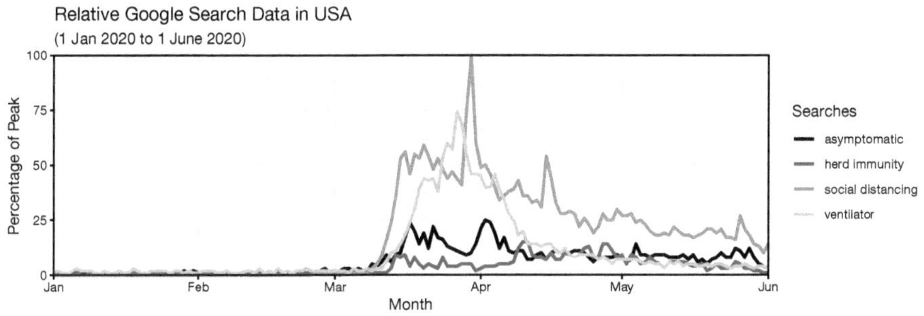

Fig. 1.3 Google Trends data in USA showing the same data as Fig. 1.2 without normalization. All percentages are relative to the peak of "social distancing" searches in late March. Data sourced from https://trends.google.com/trends/. Graph by author in R using ggplot2 (Wickham, 2016)

The propagation of concepts and language related to COVID-19 did not go as smoothly as it could have. Not only was there considerable confusion among the public about the nature of the virus and the correct response, but inter-organizational disagreements over the language of COVID-19 spilled into the public view. For one, people in positions of authority could not agree on what to call the virus. While officially named "SARS-CoV-2" in February 2020 by the International Committee on Taxonomy of Viruses (ICTV), this name for the virus was avoided by the WHO and other organizations because of the worry that people would think SARS-CoV-2 was the same virus responsible for the 2002–2004 SARS outbreak (Enserink, 2020). This worry was anticipated in the article that initially coined "SARS-CoV-2", where the authors go to some length to defend the name while explicitly distancing it from SARS (Gorbalenya et al., 2020, pp. 537–539). The WHO was not convinced by the arguments of the ICTV and elected not to call the virus "SARS-CoV-2". Instead, the WHO announced they would call the *virus* "COVID-19", even though "COVID-19" was by that time the accepted name of the *disease* caused by the virus (Enserink, 2020).[2] Moreover, despite the WHO's status as the premier global public health body as well as the preference of many news organizations to use "COVID-19", Google Trends data indicates that for the first few months of the pandemic, "coronavirus" was a far more popular name than "SARS-CoV-2", "COVID-19", and "COVID"—although by the end of May 2020, "COVID" overtook and eventually almost completely replaced "coronavirus" ('Google Trends', 2022).[3] Therefore, despite competing efforts of the ICTV and the WHO to control the vocabulary of the virus, everyday people adopted a different name altogether at the start of the pandemic.

"Social distancing" also stuck despite efforts by governments, NGOs, and public figures to push other language. The troubled introduction of social distancing mea-

[2] This chapter will follow the convention of distinguishing SARS-CoV-2 (the virus) from COVID-19 (the disease) due to considerations of expressive power.

[3] All observations of use were made up to the time of resubmission, December 2022.

sures, which included packed beaches and bars (Nelson, 2020; Phillips & Jungreis, 2020; Schnell, 2020), led many to conclude that the label "social distancing" was partially to blame. This culminated around mid-March 2020, where in the span of around 2 weeks, the phrase was both criticized publicly and abandoned by many institutions. Critical of "social distancing", the social psychologist Regan Gurung told the *USA Today* in an article published on 17 March 2020,

> If you start practicing social distancing—*which we really need to call physical distancing so people better understand the concept*—it highlights the fact that, oh my gosh, this is a dangerous time. (Schnell, 2020) [emphasis added]

That same week, psychologist Jamil Zaki was widely quoted as saying, "We should think of this time as 'physical distancing' to emphasize that we can remain socially connected even while being apart" and suggested people use "distant socializing" instead (De Witte, 2020). Taking action on 20 March 2020, the WHO publicly adopted "physical distancing" as their official language (Azis, 2020), a move that was subsequently copied by many governments and news organizations (e.g., Canada, 2020; R. Davies, 2020). Nonetheless, "social distancing" remained the common name for social distancing measures (Sørensen et al., 2021).

The debates that spilled into the public eye over the vocabulary of the pandemic are, at their heart, debates over whether "social distancing", "coronavirus", "SARS-CoV-2", etc., are good labels for the corresponding concepts, meanings, and referents. This chapter develops a framework for understanding what makes a good label and what corresponding lessons can be learned by conceptual engineers. By taking into account the three-way causal relationship between terms, mental representations, and the non-linguistic, non-mental world, the framework can be used to develop desiderata for what language to use when propagating concepts or meanings. I argue particular focus needs to paid to the non-semantic and non-pragmatic information carried by a word or phrase itself and the *lexical effects* (Cappelen, 2018, Chapter 11) such information can have on the mental representations of language users. By focusing on such lexical effects and the goals of propagation, both conceptual engineers and specialists in high-impact fields like epidemiology can better engineer their language for propagation.

Section 1.2 discusses how the form words or phrases take (that is, their phonology and graphology) can carry information that can lead to both linguistic and non-linguistic knowledge and develops a three-part framework for understanding this process. Section 1.3 discusses this framework in relationship to the goals of a public health response to a pandemic. Sections 1.4 and 1.5 apply this framework to "coronavirus" and "social distancing", respectively. I ultimately endorse "coronavirus" as a mostly unproblematic name for SARS-CoV-2 but argue that using "social distancing" as the name for social distancing measures was a mistake almost two decades in the making.

1.2 How Terms Teach and Mislead

In this section, I develop a framework for understanding what makes a good or bad name for a meaning or concept by focusing on the information carried by a name's appearance. Good names help listeners (or readers) develop true mental representations by teaching a listener its meaning, helping form a concept (or conceptual competence), and/or teaching the listener something about the world. Bad or maliciously chosen names will instead help cause false beliefs or propagate defective concepts.

When discussing words and phrases, philosophers of language have typically focused on their semantic and pragmatic meaning. Discussion in this context revolves around what a word or phrase means, either as a type or as a token utterance, and how it comes to have that meaning. There is nonetheless more to language than semantic content, pragmatic content, truth values, etc. Words and phrases also have non-semantic, non-pragmatic causal power, or *lexical effects* (Cappelen, 2018, Chapter 11).[4] To consider a lexical effect related to the pandemic, consider the politicized use of "Wuhan Virus". Semantically, the name refers to SARS-CoV-2 and, depending on how we understand the semantics of capitalized descriptions (see Rabern, 2015), means either *the virus from Wuhan* or *SARS-CoV-2*. Depending on the context, the name might also have pragmatic force. The choice to use "Wuhan Virus" in conversation, given the other options available to talk about the virus, may signal via Gricean mechanisms certain content about political affiliation. Semantic and pragmatic content aside, there are other things "Wuhan Virus" can accomplish, and these are its lexical effects. Most obviously, and as savvy political operators surely knew, using "Wuhan Virus" instead of more sterile language builds associations in people's heads between China and the pandemic. Indeed, the lexical effects might even go a step further. Someone ignorant of SARS-CoV-2's origin may, upon hearing the sentence "Nadja caught the Wuhan Virus" form *true non-linguistic beliefs* based on the name "the Wuhan Virus" that SARS-CoV-2's origin involved Wuhan.

The idea of inferring worldly facts from names might seem at odds with the standard view that the connection between a word's form and its semantic meaning and referent is arbitrary. In principle, any string of letters or series of sounds can have any meaning. Nonetheless, there are regularities between the appearance of words and their semantic value that language learners can and do exploit. These connections include *iconicity*—form-meaning resemblance—and *transparency*—the extent to which the meaning of a multimorphemic word is interpretable from the way its component parts combine. Words like "whoosh" or "zigzag" resemble whooshes and zigzags and are therefore iconic (Dingemanse

[4] This is notion is inspired by Frege's discussion of "colouring" (Frege, 1997; Sander, 2019). Following Cappelen, I am understanding the semantic/pragmatic distinction as Gricean. For a discussion of similar phenomena through the lens of relevance theory instead of Gricean pragmatics, see Carston (2019)

et al., 2015; Lupyan & Winter, 2018). Words like "campervan" (*camper + van*) and "flammable" (*flamm + able*) mean what their parts suggest they mean and are therefore transparent (Libben et al., 2003). These regularities are not always as obvious or apparently principled as transparency or iconicity, though. For example, English nouns are on average longer than English verbs (Cassidy & Kelly, 1991, 2001), and velar plosives ([k] and [g]) are disproportionately common in English slurs (Mandelbaum et al., 2024).

Mature language users use information like iconicity, transparency, and other regularities to interpret semantic and non-semantic facts (Libben, 1998; Mandelbaum et al., 2024; To et al., 2016). For example, despite not being an established word, "unflabbergastable" has a clear meaning to you as a reader because of regularities in English word construction. In order to talk about this interpretive process, I will talk about what I call a term's *decipherability*:

> The degree to which someone is apt to infer *accurate* information from a term itself, given their other mental representations.

Decipherability involves three players: *the term* that serves as a label and linguistic vehicle; a person's *mental representations,* which affect how the term is interpreted and which are affected by the term; and *the world* external to the person's psychological states, which is both represented by the person and affected by the person's mental representations.[5]

The rest of the chapter will use these categories to capture the pre-theoretic distinction between what a word or phrase looks like—*the term*—what is in the head—*the mental representations*—and what is not in the head—*the world.* These are meant to be theoretically-neutral metaphysical joints that any philosophical framework will acknowledge in one way or another: there are the visual and/or acoustic forms taken by linguistic entities, there are things "in the head" whose existence is grounded in our own cognitive states, and there are things "outside the head" whose existence is not grounded on our mental representations. Because these are meant to be pretheoretic categories, I will take pains to not take substantive commitments about what goes into each category, instead allowing the reader to fill in the categories as they wish. For example, Fregeans and other externalists about concepts will generally place concepts in the *world* category whereas internalists will generally place concepts in the *mental representation* category. Similar differences will exist between semantic internalist and externalists about meaning as well as nominalists and platonists about kinds and other purported abstracta.

The same interpretive process that allows us to understand the meaning of "unflabbergastable" without context can backfire. People can form mental repre-

[5] This 3-part distinction, as well as discussion below of the causal relations between each part, is inspired by discussion of the semiotic triangle in Ogden and Richards (1930). Unlike Ogden and Richards, who, in current philosophical jargon, are probably best described as semantic and content internalists, discussion here is meant to capture pre-theoretic joints that can accommodate various forms of externalism as well as internalism.

sentations based on a term that are inaccurate. Indeed, the names given to pathogens and diseases have a long history of these interpretive processes misfiring. During the 2009 swine flu pandemic, which was named "swine flu" because the virus crossed over to humans from pigs, pork sales dropped across the world despite cooked pork being safe to eat. After the 2009 pandemic, ABC News interviewed an American pork farmer, who said:

> We were about to turn the corner and start making a profit [. . .] And here somebody labeled H1N1 the 'swine flu,' and it just totally took a nosedive. (Selyukh, 2009)

By the farmer's telling, the name "swine flu" lead people to form false mental representations that pork was in some way unsafe or unsanitary. Similarly, the inclusion of place names in the common names of diseases or pathogens (such as the Middle East Respiratory Syndrome—MERS—or the Spanish Flu) has caused stigmatization and even attacks on people from the area. Recognizing this, WHO guidelines recommend against naming diseases after places, animals, industries, or people. Instead the WHO recommends naming diseases after things such as the bodily systems affected, the year of discovery, symptoms, and the disease's level of severity (World Health Organization, 2015).

Because the same processes involved with decipherability can lead to inaccurate mental representations, any discussion of decipherability needs the opposing notion of what I will call *misleadingness*:

> The degree to which someone is disposed to infer *inaccurate* information from the term itself, given their other mental representations.

Misleading terms lead to mental representations—whether beliefs about the world, mental imagery, concepts (understood here as token cognitive entities), or metalinguistic beliefs about the terms—that are inaccurate, whereas decipherable terms lead to accurate mental representations. As will be important for the language of COVID-19, the misleadingness or decipherability of a term is individual-specific. The exoplanet *K2-198d* or the anatomical part *antecubital fossa* have names that most readers cannot extract information from even though the terms are decipherable to astronomers and people knowledgeable of Latin, respectively. Correspondingly, a term that works well in one community may not work well in another, especially when there is a large knowledge, cultural, or conceptual gap between the two.[6]

1.3 What we Want from Epidemiological Language

When an emergency necessitates the rapid propagation of ideas, beliefs, and concepts, decipherability is an essential tool in the toolbox of politicians, journalists,

[6] For this reason, the conclusions drawn below about people's reaction to "coronavirus" and "social distancing" should be seen as being limited to fluent English speakers without expertise in epidemiology, public health, or other relevant fields.

and academics. Where decipherable terms will help save lives by helping inform language-users, misleading terms will do the opposite. Creating a decipherable term is easier said than done however, because a term's decipherability is limited by the tension between the amount of encoded information and the usability of a term. In this section, I examine what information should be encoded in epidemiological language so that we can evaluate "coronavirus" and "social distancing" accordingly.

As discussed above, designing decipherable language is a useful way to propagate beliefs and concepts, but decipherability is not the only consideration that determines the success of propagation. Choices need to be made about what information a term should carry because the more information a term carries, the less easy it is for language users to use. For example, if we wanted to give COVID-19 a name that maximized information, we might call it "The Disease Caused By SARS-CoV-2 That Spreads Through Air, Causes Coughs, And Sometimes Leads to Long COVID". However, it will be hard to get language users to adopt this name. Compared to "COVID", the name is clunky, hard to remember, and time-consuming to write or say. Therefore, when choosing the term to propagate, our goals for decipherability need to be modest and targeted.

To this end, the goal of an epidemiological response, from the earliest stages of an outbreak response to the final stages of a pandemic, is to limit the damage of the disease (World Health Organization, 2018, p. 1). What steps are taken to limit the disease's impact are different at different stages of the outbreak, but a large part of an outbreak response, and the focus of the rest of this chapter, is managing people's behavior. While some behavior can be forced—people cannot enter a country by commercial flights if commercial flights are cancelled—other behavior needs to be taught. Healthcare communication is one such way to manage learned behavior to limit the damage of a disease. For this reason, the goals of introducing a term for non-specialist consumption in public health emergencies should prioritize true beliefs about disease mitigation over true beliefs about the disease itself.

Returning to last section's framework of decipherability and misleadingness, this gives us a desired design outcome for decipherable pandemic-related language. In the *world*, we want to limit damage of the disease, using people's behavior to do so. As a means for this, public health responses need to propagate *mental representations* that lead good behavioral and epidemiological outcomes. *Terms* are successfully decipherable in this instance insofar as people can infer information from the term that leads to damage reduction. Importantly, in this instance such information is both doxastic and conceptual. Not only are people learning the meaning of terms when they learn "coronavirus" and "social distance", but they are also learning new ways of categorizing the world. Prior to 2020, many of us lacked the ability to believe things like *social distancing is a way to reduce the spread of disease* because we lacked a constituent of that belief, namely SOCIAL DISTANCING. Therefore, the decipherability or misleadingness of "coronavirus" and "social distancing" goes beyond merely the process of learning the meaning of phrases but includes conceptual propagation as well. Now that everything is in place, it is time to turn to "coronavirus" and "social distancing".

1.4 Case Study 1: "Coronavirus"

When examined through the framework of decipherability and misleadingness, "coronavirus" is a good term for the virus—with one notable exception. The ways in which it is good and bad are illustrative, and we can learn about the challenges facing people who want or need to propagate words or concepts. This section starts by looking at the contrast between "coronavirus" and "proper" scientific terminology. For one, "coronavirus" is a loose metonymic name for a species of the virological family by the same name. In addition, "coronavirus" was used to describe both the virus and disease despite official channels trying to introduce "SARS-CoV-2" and "COVID-19" to contrast the two. I argue that only the metonymic shift is a problem, and it is only a problem because an earlier metonymic shift of "coronavirus" made the term misleading to some people. Finally, I look at the way the meaning of "coronavirus" is partially decipherable. While it is clear the referent is a virus, it is not clear to those without microbiological knowledge how "corona" modifies virus. This is, if anything, a strength of the term, as it presents important information while remaining future-proof.

Using "coronavirus" as a catch-all term for the outbreak is not a scientifically rigorous way of speaking—it is a metonymic use of the virus species' family name. Biologically speaking, calling SARS-CoV-2 "coronavirus" or "the coronavirus" is roughly the same as referring to house cats as "feline" or "the feline". We might worry this metonymic looseness is itself a problem or even that using "coronavirus" as a name for SARS-CoV-2 propagates false semantic beliefs about the referent of "coronavirus".[7] However, as discussed in Sect. 1.3, the key desiderata of a public health response is limiting damage of a disease, so the semantic or metonymic looseness of a term is only problematic to those ends.

The metonymic shift from family to species *did,* however, hurt the pandemic response. The shift in 2020 was not the first time "coronavirus" metonymically shifted to a subset of the family. "Coronavirus" had been used prior to 2020 as a name for the coronaviruses that cause disease in humans (OED3, 2022). Because of this older use, some packages of disinfectant written before spring 2020 included "coronavirus" or "human coronavirus" on the list of bacteria and viruses that the disinfectant killed. This fueled conspiracy theories that governments have known about SARS-CoV-2 for years (Chalmers, 2020).[8] In this way, "coronavirus" was a misleading label to some because they failed to recognize that "coronavirus" had

[7] See especially classic semantic externalist and content externalist arguments about natural kind terms (Burge, 1979; Kripke, 1981; Putnam, 1975). See (LaPorte, 2003; Sawyer, 2020; Sterken, 2020) for contemporary discussion of these arguments and (Wikforss, 2007) for an introduction to the debate.

[8] The corresponding disputes over statements like "Coronavirus arose in 2019" are verbal disputes, although different in form from stereotypical examples (Belleri, 2018; Inga, 2018; see Jenkins, 2014). Conspiracy theorists assert sentence P because they believe "coronavirus" has semantic value X, where public health officials assert sentence P is false because they believe "coronavirus" has semantic values X *and* Y.

multiple related uses. Nonetheless, without empirical study beyond the scope of this chapter, it is impossible to say how many people were misled by pre-COVID uses of "coronavirus". Confusion may have been limited to a few people on social media already prone to conspiracy theorizing, or it may have had a larger effect on people's response to the pandemic.[9]

Another potential problem with "coronavirus" is that it runs against the ICTV's designation of "SARS-CoV-2", which was introduced in part to allow the distinction between the virus and the disease caused by the virus (Gorbalenya et al., 2020). There is indeed some evidence that the failure to follow naming protocols helped enable conflation of the disease and the virus. For example, consider this line from a news report from *Reuters* in March 2020:

> A second member of France's National Assembly has been taken to hospital after contracting *coronavirus* and five other lawmakers are being tested for the *illness* [emphasis added] (Rosemain, 2020)

Given that "the illness" is anaphorically referring to "coronavirus", it appears "coronavirus" is intended to refer to the disease caused by the virus, not the virus itself. There does not seem to be evidence that such conflation was problematically misleading, however. English has the tools to disambiguate the virus and the disease when needed. For one, "COVID" and "COVID-19"—which are shortened forms of COronaVIrus Disease 2019—were used early on to specifically refer to the disease (see OED Editorial Board, 2020). For example, in May 2020 the following line in a different *Reuters* article uses "COVID-19" to distinguish the illness from the virus:

> Bright [...] said he was ousted from BARDA because he resisted efforts to push hydroxychloroquine and the related chloroquine as cures for COVID-19, *the respiratory illness caused by the coronavirus*. [emphasis added] (Wolfe, 2020)

While many circumstances do not require disambiguating the virus from the disease (little is gained or lost if a sign reads "coronavirus testing center" compared to "COVID-19 testing center"), we have these different names when necessary. Moreover, beyond using "COVID-19", other strategies exist to draw the distinction, such as "coronavirus illness", "coronavirus infections", or describing someone as "ill from coronavirus". While perhaps this is not as clear of a distinction as those responsible for the name "SARS-CoV-2" were hoping for, English has ample expressive power to avoid confusion.

Speaking in the term's favor, "coronavirus" is decipherable, albeit to a limited extent. "Coronavirus" is a compound noun composed of two morphemes, *corona* and *virus*. Both are well-known roots although some people may struggle to define either term when they first hear the compound noun. In addition, despite being an adjective-noun phrases, only those with certain microbiological knowledge will be able to decipher that *corona* modifies *virus* in describing the visual appearance

[9] "Coronavirus" was not the only name of the virus/disease that enabled conspiracy theorizing. Multiple prominent public figures inferred from "COVID-19" that the disease was the 19th COVID (Aodha, 2020; Lee, 2021).

of the virus under a microscope. Notably, because the scientific understanding of SARS-CoV-2 evolved considerably over the course of the pandemic, it is probably for the best that public figures did not aim for something more decipherable than "coronavirus". In early 2020 when specialists needed to introduce a term into the public's vocabulary to discuss the virus and mitigation factors, many features of the virus were not yet understood. For example, while early research found evidence that SARS-CoV-2 could remain alive on surfaces for days at a time—spreading fear of surface-based infection—consensus was eventually reached that surface-based transmission was far less common than airborne transmission (see CDC, 2021). A term that aimed to carry more information, such as one that was apt to cause mental representations that SARS-CoV-2 was transmitted primarily through surfaces rather than primarily through the air, risked turning out to be misleading rather than decipherable.

1.4.1 Lessons from "Coronavirus"

Given the positives and negatives of "coronavirus", it is worth pausing to think about what people in a position to propagate new vocabulary—whether conceptual engineers or public health officials—can learn from "coronavirus". First, the unimportance of following technical naming conventions for the sake of following technical naming conventions, at least for the purposes of public health responses, means that propagators need to reflect on the goals they have for propagation (see Isaac et al. (2022) for a discussion of the relevant distinction between the goals and targets of conceptual engineering). Different goals will result in different desiderata for propagation. In this case, the goal of a pandemic response is to limit the damage of a disease, so semantic accuracy or adherence to specialist norms takes a backseat to spreading safe behavior.

Second, the conspiracy theorizing enabled by competing metonymic uses of "coronavirus" suggests propagators should pay attention to whether propagating a term will cause problematic polysemy. As we saw with "coronavirus", introducing a metonymic use of a term can bring about unwanted inferences caused by a different uses of the term.

Third, propagators should future-proof their language. As was the case with SARS-CoV-2's method of transmission, we may have incomplete understanding of a referent at the time of propagation. A propagated term should therefore, as much as is possible, be insulated against future changes in beliefs or understanding.

1.5 Case Study 2: "Social Distancing"

"Social distancing" shares a number of features with "coronavirus". Both are multi-morphemic nouns composed of an adjective-noun pairing that before the 2020 pandemic were used almost exclusively in technical settings. Nevertheless, the two terms draw an illustrative contrast. Whereas the framework thus discussed gives mixed-to-positive results to the suitability of "coronavirus" as a label for SARS-CoV-2, "social distancing" appears to be a straightforwardly bad name for social distancing. Moreover, when we look at the historically contingent reasons that seem to be behind the propagation of "social distancing" in early 2020, we are pushed to think carefully about the way the vocabulary specialists use in communication with each other can affect efforts at conceptual and linguistic propagation.

The term "social distancing" appears to have first entered epidemiology in 2005 (Ferguson et al., 2005). While the strategy of social distancing is itself not new (Sørensen et al., 2021)—leper colonies are, after all, a form of social distancing—the explicit epidemiological strategy of "increasing social distance" owes its origin to scholarly discussion of the 2002–2004 SARS outbreak (see Bell, 2004; Gostin, 2004; World Health Organization, 2005). From the coining of the term onward, epidemiologists used "social distancing" and "increasing social distance" as terms of art to describe strategies of "reducing contact rates in the population" (Ferguson et al., 2005, p. 209). This usually included the measures of closing schools, banning large public gatherings, keeping non-essential workers home, and voluntary self-isolation of people with symptoms (Ahmed et al., 2018; Glass et al., 2006; Herrera-Valdez et al., 2011, p. 24; e.g., World Health Organization, 2005, p. 59). Research found that social distancing measures were generally effective in reducing the spread of viral diseases, and it was adopted as a strategy by epidemiologists and other public health officials. Despite this, "social distancing" did not enter everyday language because social distancing measures, while seeing limited use during the 2009 swine flu pandemic (Herrera-Valdez et al., 2011; Horney et al., 2010), were not widely used between the terms' introduction in 2005 and the COVID-19 pandemic in 2020.

Because social distancing was unknown to most people, when public health figures worldwide decided that social distancing was an appropriate response to the 2020 COVID-19 pandemic, social distancing had to be explained to the public. During this stage, the notion of *social distancing* appears to have been simplified slightly. Understandably, few public figures followed the epidemiological literature in defining "social distancing" in terms of contact rates. Social distancing was instead explained as types of actions individuals can take, but the spirit of the term's original definition stuck. The BBC explained social distancing as "no coming in contact with other people unless you need to" (*Social distancing* 2020). The CDC similarly said social distancing "means keeping space between yourself and other people outside of your home" (CDC, 2020), and the WHO explained it as maintaining 1 meter or more distance between you and people coughing or sneezing (World Health Organization, 2020).

Regardless of whether we understand "social distancing" in its original technical sense to mean *reducing contact rates in the population* or the more individual-specific *not coming into contact with other people unless you have to*, "social distancing" is misleading. The term carries inaccurate information about both the meaning of the term and what the strategy of social distancing involves. When hundreds of millions of English speakers heard of social distancing for the first time during the early months of the COVID-19 pandemic, they had to try to figure out what exactly "social distancing" meant. Given it is an adjective-noun pairing, the most natural interpretation is that social distancing is a process by which we distance ourselves socially (Presterudstuen, 2020). Whereas distancing oneself socially covers a lot of behavior that *will* reduce transmission—physically meeting up with friends, going to a bar, attending parties—much is missed. Many spaces that were thought early in the pandemic to carry high risks of transmission are not spaces that are typically thought of as social spaces, such as grocery stores, airplanes, and hospitals. Similarly, going alone to a crowded indoor space is a way of being socially distanced without *social distancing*. On the flip side, distancing oneself socially excludes safe social distancing behaviors, such as carrying on a conversation across an empty street.

As with the earlier question of how much conspiracy theories were driven by older use of "coronavirus", it is not the philosopher's place to say how exactly the misleadingness of "social distancing" played out. This is best left to sociocultural linguists and others with the methods capable of examining language's effect on behavior at the population-level. Nevertheless, bad social distancing behavior suggests that the misleadingness adversely influenced peoples' mental representations about social distancing. At the very start of the pandemic, owners of remote parking lots in the Scottish Highlands had to chase off people in campervans planning to wait out the pandemic (Brooks, 2020). While hiding in a campervan is an effective way of distancing oneself socially, because of the risk of bringing the disease to a remote area with a correspondingly low healthcare capacity, this was not effective social distancing. Making the opposite mistake, when local police in Mississippi ticketed people in cars at a drive-in church service broadcast to participants by FM radio (Williams, 2020), the local government punished an effective social distancing measure where people were not distancing themselves socially. While participants were engaging in the social practice of a church service, they were keeping contact rates low by keeping space between themselves and people outside of their household. Therefore, punishing the church goers in this circumstance is justified if we understand social distancing as *distancing that is social*, but it is not justified if we understand social distancing as any of the definitions for social distancing quoted above.

A failure to recognize or address the misleading nature of "social distancing" may have exacerbated problems. Many news organizations and governments used problematic definitions like "reducing social *interactions* between people" or "limiting social *activities*" [emphasis added] (Mazziotta 2020; e.g., Nelson, 2020; Welsh Government 2020). Definitions such as these do not clarify that "social distancing" is not about distancing that is social but about behavioral steps that reduce the reproductive rate of the virus by limiting physical proximity. This was a limited problem though, as many early explanations of the strategy involved both an intensional definition (e.g., avoid physical contact that might spread the disease) and extensional examples (e.g., hold meetings remotely, only go shopping for essentials) that help clear up potential confusion (e.g., BBC, 2020; CDC, 2020).

Despite the misleadingness, "social distancing" does have merits. Like "coronavirus", "social distancing" provided people and public figures with vital expressive power to help communicate necessary and drastic changes to day-to-day life. Moreover, the term is suitable for people with limited English vocabulary because "social" and "distance" are among the 2000 most common English words—a threshold believed to be important for understanding among people with lower levels of literacy (Rakedzon et al., 2017). Relatedly, as it consists of five syllables across two common English words, "social distancing" is useable and easy to remember. These three features are all helpful for the purposes of the public health response. However, while "social distancing" was propagated to help instill certain behavior that slows the spread of infectious diseases, "social distancing" is not the only term that can fulfill these criteria. The widely recommended "physical distancing" is also usable and easy to remember, offers expressive power, and only uses common English words. While the Corpus of Contemporary American English (COCA) shows that since 1991, "physical" is about a third as common in American English as "social" (M. Davies, 2010), "physical" is still common enough to be suitable for low-literacy English speakers (Rakedzon et al., 2017).

How do "social distancing" and "physical distancing" *differ*? Unlike "social distancing", "physical distancing" does not mislead people into thinking they should be distancing themselves socially. There are plausibly situations where one could social distance effectively while being in close physical proximity to other people—such as with the careful use of high-grade personal protective equipment (PPE)—but, at the start of the pandemic, such opportunities were few and far between. Nonetheless, "social distancing" is apparently better than "physical distancing" in at least one way. "Social distancing" adheres to pre-existing epidemiological vocabulary. As discussed in relationship to "coronavirus", however, using technical language properly is less important during a pandemic than preventing people from getting sick. Therefore, since "social distancing" offers the large and unique drawback of being misleading in a way that potentially led to higher rates of infection, "physical distancing" should have been the language propagated from the start.

1.5.1 Lessons from "Social Distancing"

While the misleadingness of "social distancing" is itself problematic, "social distancing" offers a larger cautionary tale for people in a position to propagate new terms—whether conceptual engineers, scientific experts, or public figures. "Social distancing" appears to have been the term used by public figures because that was the term epidemiologists had been using to refer to social distancing, not because "social distancing" was the best term to propagate to non-experts. For this reason, "social distancing" invites those thinking about propagating a term, or even those just using a term that might be propagated, to reflect on how the language they use might affect the population at large.

As discussed above, "social distancing" was first used by epidemiologists to discuss the strategy of closing schools, workplaces, and large venues such as concert halls during 2002–2004 SARS outbreak Southeast Asia. The choice to call it "social distancing" instead of "physical distancing" was appropriate at the time. As Skype was only founded in 2003 and the first iPhone would not be sold until 2007, this was a time and place where many social interactions had to be physical interactions, and so the physical could be conflated with the social. The period between 2004 and 2020, however saw a revolutionary shift in how much of our lives we can and do live online. In the language of Floridi (2007), technology *reontologized* our work and social lives by blurring the line between *here* vs *there* until physical collocation was no longer an intrinsic feature of many of our day-to-day activities. Compared to 2004, in 2020 it was far easier for many to edit company documents without being on a corporate local network, rent a movie without a trip to a rental store, and attend class synchronously over webcam. Therefore, as technology progressed between 2004 and 2020, "social distancing" became more misleading as it was more and more possible to lead a social life while physically removed from others.[10]

As is common in academic writing, language was slow to change, and specialists continued using "social distancing" even as it grew more and more misleading. This was not in itself a problem until "social distancing" entered everyday language. From its coining in 2005 until early 2020, "social distancing" was largely kept to epidemiological circles. This meant that "social distancing" was learned by high-literacy people who had the time, motivation, and expertise to work past the misleadingness of the term. Potentially complicating matters, it is possible specialists did not even notice "social distancing" was misleading. Technical language is insular in nature, and it is easy to lose track of which of our concepts, words, and beliefs are due to our expertise. However, when a pandemic forced fast dissemination of knowledge, "social distancing" was the term public figures were left with for social distancing, and the term and idea behind the term had to be learned by people who did not have the same time or specialized knowledge as experts. This forced public figures to react to the misleadingness of "social distancing" after the misleading term had already been propagated.

In fields like epidemiology, members of the discipline do not have control over when a term is propagated, so avoiding a mistake like "social distancing" requires looking inward. If epidemiologists do not know when a notion as important as

[10] Another possibility is that "social distancing" became semantically *demotivated*, where change, whether semantic or worldly, leads to a transparent term losing its transparency—that is, it becomes *opaque* (Blank, 2001). For example, American football has the same foot-based origin as rugby and soccer, but in the nineteenth century a series of rule changes led to the modern hand-based sport, leading to the demotivation of "football" in American English. On this picture of semantic demotivation, "social distancing" became demotivated as its meaning shifted from targeting social events to reducing contact rates more generally. However, the earliest epidemiological discussions of increasing "social distance" already include strategies like the use of facemasks (Bell, 2004, p. 1902; World Health Organization, 2005, p. 58) and the cancellation of public transit (Gostin, 2004, p. 570). This suggests "social distancing" did not demotivate because it was partially opaque from the start.

social distancing will need to be propagated, then a term that is decipherable to folk needs to be used *even in communication between experts*. The language used in such communication is, after all, the language that public figures such as politicians and journalists will reach for during a crisis. Therefore, specialists in any field whose discoveries may 1 day be of public concern should act now to clean up their language instead of waiting until the horse has already left the barn.

Conceptual engineers in contrast have some control over when a term is propagated. Therefore, avoiding a situation like "social distancing" requires looking outward. Conceptual engineers need to test the language they use for a concept or meaning before they propagate it. We might not be able to predict with certainty the future route concepts or meanings will take (Marques, 2020), but not everything is unknowable. We can make generalizations based on past patterns of concept and meaning change (Koslow, 2022; Thomasson, 2021), and time will tell whether conceptual engineers can develop an empirical framework for beta testing terms, meanings, and concepts for propagation.

1.6 Conclusion

This chapter has looked at the lessons that conceptual engineers can draw from the early days of the COVID-19 pandemic. The discussion focused on the way language carries epistemic information that can inform and mislead language users. Decipherable language carries information that can aid in the acquisition of accurate semantic beliefs, object-level beliefs, or even the formation of new concepts or conceptual competence whereas misleading language gives rise to inaccurate beliefs or defective concepts. "Coronavirus" is easily rememberable but misleading insofar as a previous use of "coronavirus" enabled conspiratorial beliefs about the origin of SARS-CoV-2. "Social distancing" is a misleading name for social distancing measures, and it ended up in circulation because epidemiologists had been using the term without issue for years. While this chapter has only scratched the surface of one of the many issues facing conceptual and linguistic propagation, hopefully conceptual engineers can continue to learn how to harness propagation by looking at the many examples of propagation of scientific language into everyday use.

References

Ahmed, F., Zviedrite, N., & Uzicanin, A. (2018). Effectiveness of workplace social distancing measures in reducing influenza transmission: A systematic review. *BMC Public Health, 18*(1), 518. https://doi.org/10.1186/s12889-018-5446-1

Aodha, G. N. (2020, April 22). Explainer: No, 'Covid-19' doesn't mean there have been 18 other coronaviruses. *TheJournal.ie*. https://www.thejournal.ie/factcheck-covid-19-other-coronaviruses-5081369-Apr2020/. Accessed 15 August 2022

Azis, S. (2020, March 30). Why 'physical distancing' is better than 'social distancing'. *Al Jazeera*. https://www.aljazeera.com/news/2020/03/physical-distancing-social-distancing-200330143325112.html. Accessed 15 May 2020
BBC. (2020, February 11). Coronavirus disease named Covid-19. *BBC News*. https://www.bbc.com/news/world-asia-china-51466362. Accessed 9 April 2020
Bell, D. M. (2004). Public health interventions and SARS Spread, 2003. *Emerging Infectious Diseases, 10*(11), 1900–1906. https://doi.org/10.3201/eid1011.040729
Belleri, D. (2018). Two species of merely verbal disputes. *Metaphilosophy, 49*(5), 691–710. https://doi.org/10.1111/meta.12335
Blank, A. (2001). *Pathways of lexicalization. Language Typology and Language Universals*. De Gruyter Mouton. https://doi.org/10.1515/9783110194265-049
Brooks, L. (2020, March 22). Scottish Government Furious at Travellers to Highlands and Islands. *The Guardian*. https://www.theguardian.com/world/2020/mar/21/scottish-government-furious-at-travellers-to-highlands-and-islands. Accessed 15 April 2020
Burge, T. (1979). Individualism and the mental. *Midwest Studies In Philosophy, 4*(1), 73–121. https://doi.org/10.1111/j.1475-4975.1979.tb00374.x
Canada, P. H. A. of. (2020, April 15). Physical distancing: How to slow the spread of COVID-19. *AEM* Education and Awareness. https://www.canada.ca/en/public-health/services/publications/diseases-conditions/social-distancing.html. Accessed 16 April 2020
Cappelen, H. (2018). *Fixing language*. Oxford University Press. https://doi.org/10.1093/oso/9780198814719.001.0001
Carston, R. (2019). Ad hoc concepts, polysemy and the lexicon. In B. Clark, K. Scott, & R. Carston (Eds.), *Relevance, pragmatics and interpretation* (pp. 150–162). Cambridge University Press. https://doi.org/10.1017/9781108290593.014
Cassidy, K. W., & Kelly, M. H. (1991). Phonological information for grammatical category assignments. *Journal of Memory and Language, 30*(3), 348–369. https://doi.org/10.1016/0749-596X(91)90041-H
Cassidy, K. W., & Kelly, M. H. (2001). Children's use of phonology to infer grammatical class in vocabulary learning. *Psychonomic Bulletin & Review, 8*(3), 519–523.
CDC. (2020, February 11). *Social distancing, quarantining, and isolation*. Centers for Disease Control and Prevention. https://www.cdc.gov/coronavirus/2019-ncov/prevent-getting-sick/social-distancing.html. Accessed 15 April 2020
CDC. (2021, March 24). *Science brief: SARS-CoV-2 and surface (fomite) transmission for indoor community environments*. Centers for Disease Control and Prevention. https://www.cdc.gov/coronavirus/2019-ncov/more/science-and-research/surface-transmission.html. Accessed 15 August 2022
Chalmers, V. (2020, January 31). Dettol CAN'T kill the deadly Wuhan coronavirus that has reached UK. *Mail Online*. https://www.dailymail.co.uk/health/article-7951687/Dettol-kill-deadly-Wuhan-coronavirus-finally-reached-UK.html. Accessed 29 April 2020
Coronavirus: What are the rules for social distancing? (2020). BBC News. https://www.bbc.co.uk/news/av/health-51966112/coronavirus-what-are-the-rules-for-social-distancing. Accessed 10 April 2020
Davies, M. (2010). The corpus of contemporary American English as the first reliable monitor corpus of English. *Literary and Linguistic Computing, 25*(4), 447–464.
Davies, R. (2020, May 12). Physical distancing may extend to 2022, COVID-19 study finds. *The Guardian*. https://www.theguardian.com/uk-news/2020/may/12/physical-distancing-public-transport-dft-government-advice-lockdown. Accessed 15 May 2020
De Witte, M. (2020, March 19). Try 'distant socializing' instead. *Stanford News*. https://news.stanford.edu/2020/03/19/try-distant-socializing-instead/. Accessed 31 May 2022
Dingemanse, M., Blasi, D. E., Lupyan, G., Christiansen, M. H., & Monaghan, P. (2015). Arbitrariness, iconicity, and systematicity in language. *Trends in Cognitive Sciences, 19*(10), 603–615.

Enserink, M. (2020, February 12). Update: 'A bit chaotic.' Christening of new coronavirus and its disease name create confusion. *Science | AAAS*. https://www.sciencemag.org/news/2020/02/bit-chaotic-christening-new-coronavirus-and-its-disease-name-create-confusion. Accessed 9 April 2020

Ferguson, N. M., Cummings, D. A. T., Cauchemez, S., Fraser, C., Riley, S., Meeyai, A., et al. (2005). Strategies for containing an emerging influenza pandemic in Southeast Asia. *Nature, 437*(7056), 209–214. https://doi.org/10.1038/nature04017

Floridi, L. (2007). A look into the future impact of ICT on our lives. *The Information Society, 23*(1), 59–64.

Frege, G. (1997). Logic. In M. Beaney (Ed.), P. Long & R. White (Trans.), *The Frege Reader* (pp. 227–250). Blackwell.

Glass, R. J., Glass, L. M., Beyeler, W. E., & Min, H. J. (2006). Targeted social distancing designs for pandemic influenza. *Emerging Infectious Diseases, 12*(11), 1671–1681. https://doi.org/10.3201/eid1211.060255

Google Trends. (2022, December 16). *Google Trends*. https://trends.google.com/trends/. Accessed 1 June 2022.

Gorbalenya, A. E., Baker, S. C., Baric, R. S., de Groot, R. J., Drosten, C., Gulyaeva, A. A., et al. (2020). The species severe acute respiratory syndrome related coronavirus: Classifying 2019-nCoV and naming it SARS-CoV-2. *Nature Microbiology, 5*(4), 536–544. https://doi.org/10.1038/s41564-020-0695-z

Gostin, L. O. (2004). Pandemic influenza: Public health preparedness for the next Global Health emergency. *The Journal of Law, Medicine & Ethics, 32*(4), 565–573. https://doi.org/10.1111/j.1748-720X.2004.tb01962.x

Herrera-Valdez, M. A., Cruz-Aponte, M., & Castillo-Chavez, C. (2011). Multiple outbreaks for the same pandemic: Local transportation and social distancing explain the different 'waves' of A-H1N1pdm cases observed in México during 2009. *Mathematical Biosciences & Engineering, 8*(1), 21. https://doi.org/10.3934/mbe.2011.8.21

Horney, J. A., Moore, Z., Davis, M., & MacDonald, P. D. M. (2010). Intent to receive pandemic influenza a (H1N1) vaccine, compliance with social distancing and sources of information in NC, 2009. *PLoS One, 5*(6), e11226. https://doi.org/10.1371/journal.pone.0011226

Inga, V. (2018). Verbal disputes and the varieties of Verbalness. *Erkenntnis, 83*(2), 331–348. https://doi.org/10.1007/s10670-017-9892-4

Isaac, M. G., Koch, S., & Nefdt, R. (2022). Conceptual engineering: A road map to practice. *Philosophy Compass, 17*(10), e12879.

Jenkins, C. S. I. (2014). Merely verbal disputes. *Erkenntnis, 79*(S1), 11–30.

Koslow, A. (2022). Meaning change and changing meaning. *Synthese, 200*(2), 1–26.

Kripke, S. (1981). *Naming and necessity*. Blackwell Publishers.

LaPorte, J. (2003). *Natural kinds and conceptual change*. Cambridge University Press.

Lee, B. Y. (2021, April 12). Ted Nugent mentions 18 Covids, that's not how Covid-19 got its name. *Forbes*https://www.forbes.com/sites/brucelee/2021/04/12/ted-nugent-mentions-18-covids-thats-not-how-covid-19-got-its-name/. Accessed 15 August 2022

Libben, G. (1998). Semantic transparency in the processing of compounds: Consequences for representation, processing, and impairment. *Brain and Language, 61*(1), 30–44. https://doi.org/10.1006/brln.1997.1876

Libben, G., Gibson, M., Yoon, Y. B., & Sandra, D. (2003). Compound fracture: The role of semantic transparency and morphological headedness. *Brain and Language, 84*(1), 50–64. https://doi.org/10.1016/S0093-934X(02)00520-5

Lupyan, G., & Winter, B. (2018). Language is more abstract than you think, or, why aren't languages more iconic? *Philosophical Transactions of the Royal Society B: Biological Sciences, 373*(1752), 20170137.

Mandelbaum, E., Ware, J., & Young, S. (2024). The sound of slurs: Bad sounds for bad words. In S. Nichols & J. Knobe (Eds.), *Oxford studies in experimental philosophy* (Vol. 5, pp. 221–246). Oxford University Press.

Marques, T. (2020). Amelioration vs. perversion. In T. Marques & Å. Wikforss (Eds.), *Shifting concepts: The philosophy and psychology of conceptual variability*. Oxford University Press.

Mazziotta, J. (2020, March 13). *How social distancing can reduce expected coronavirus deaths*. People.com. https://people.com/health/how-social-distancing-can-greatly-reduce-expected-coronavirus-deaths-us/. Accessed 14 April 2020.

Nelson, A. (2020, March 23). Here's why you shouldn't travel during the coronavirus pandemic, even within the UK. *The Star*. https://www.thestar.co.uk/health/it-safe-travel-around-uk-government-advice-uk-travel-after-highlands-and-lake-district-social-distancing-warnings-2506031. Accessed 3 April 2020

OED Editorial Board. (2020, July 15). Using corpora to track the language of Covid-19. *Oxford English Dictionary*. https://public.oed.com/blog/using-corpora-to-track-the-language-of-covid-19-update-2/. Accessed 31 August 2020

OED3. (2022, March). Coronavirus, n. *Oxford English dictionary online*. OED. https://oed.com/view/Entry/266178. Accessed 17 August 2022.

Ogden, C. K., & Richards, I. A. (1930). *The Meaning of Meaning*. http://archive.org/details/in.ernet.dli.2015.221615. Accessed 10 April 2020

Phillips, L., & Jungreis, M. (2020). *South Boston bars and restaurants closed Sunday as photo of line outside bar draws ire amid coronavirus outbreak*. Boston Globe. https://www.bostonglobe.com/2020/03/15/metro/photo-line-outside-boston-bar-draws-ire-establishments-be-closed-sunday/. Accessed 15 April 2020

Presterudstuen, G. H. (2020). Reclaiming the social from 'social distancing'. *Social Anthropology, 28*(2), 335–336. https://doi.org/10.1111/1469-8676.12819

Putnam, H. (1975). The meaning of 'meaning'. Minnesota *Studies in the Philosophy of Science*, 7.

Rabern, B. (2015). Descriptions which have grown capital letters. *Mind & Language, 30*(3), 292–319. https://doi.org/10.1111/mila.12081

Rakedzon, T., Segev, E., Chapnik, N., Yosef, R., & Baram-Tsabari, A. (2017). Automatic jargon identifier for scientists engaging with the public and science communication educators. *PLoS One, 12*(8), e0181742. https://doi.org/10.1371/journal.pone.0181742

Rosemain, M. (2020, March 7). Coronavirus spreads in French parliament. *Reuters*. https://uk.reuters.com/article/us-health-coronavirus-france-idUKKBN20U0RP. Accessed 14 May 2020

Sander, T. (2019). Two Misconstruals of Frege's theory of colouring. *The Philosophical Quarterly, 69*(275), 374–392. https://doi.org/10.1093/pq/pqy054

Sawyer, S. (2020). Talk and thought. In A. Burgess, H. Cappelen, & D. Plunkett (Eds.), *Conceptual engineering and conceptual ethics* (pp. 379–395). Oxford University Press.

Schnell, L. (2020, March 17). Coronavirus and social distancing: Why people won't avoid each other. *USA Today*. https://www.usatoday.com/story/news/nation/2020/03/16/coronavirus-social-distancing-why-people-wont-avoid-each-other/5065228002/. Accessed 15 April 2020

Selyukh, A. (2009, October 15). Pork Industry Still Reeling From Swine Flu. *ABC News*. https://abcnews.go.com/Business/pork-industry-reeling-swine-flu/story?id=8840004. Accessed 25 March 2020

Sørensen, K., Okan, O., Kondilis, B., & Levin-Zamir, D. (2021). Rebranding social distancing to physical distancing: Calling for a change in the health promotion vocabulary to enhance clear communication during a pandemic. *Global Health Promotion, 28*(1), 5–14. https://doi.org/10.1177/1757975920986126

Sterken, R. K. (2020). Linguistic interventions and transformative communicative disruption. In H. Cappelen, D. Plunkett, & A. Burgess (Eds.), *Conceptual engineering and conceptual ethics*. Oxford University Press.

Thomasson, A. (2021). Conceptual engineering: When do we need it? How can we do it? Inquiry, 0(0), 1–26. https://doi.org/10.1080/0020174X.2021.2000118.

To, N. L., Tighe, E. L., & Binder, K. S. (2016). Investigating morphological awareness and the processing of transparent and opaque words in adults with low literacy skills and in skilled readers. *Journal of Research in Reading, 39*(2), 171–188.

Welsh Government. (2020, March 20). *Coronavirus social distancing guidance*. GOV.WALES. https://gov.wales/coronavirus-social-distancing-guidance. Accessed 14 April 2020.

Wickham, H. (2016). *ggplot2: Elegant graphics for data analysis*. Springer. https://ggplot2.tidyverse.org

Wikforss, Å. (2007). Semantic externalism and psychological externalism: Semantic externalism and psychological externalism. *Philosophy Compass, 3*(1), 158–181. https://doi.org/10.1111/j.1747-9991.2007.00107.x

Williams, P. (2020, April 14). Justice Department defends Mississippi drive-in church service. *NBC News*. https://www.nbcnews.com/politics/justice-department/justice-department-defends-mississippi-drive-church-service-n1183761. Accessed 7 May 2020

Wolfe, J. (2020, May 14). Whistleblower to warn congress of 'grave risks' of rushing to reopen U.S. *Reuters*. https://uk.reuters.com/article/uk-health-coronavirus-usa-bright-idUKKBN22Q1JD. Accessed 14 May 2020

World Health Organization. (2005). *Avian influenza: assessing the pandemic threat* (No. WHO/CDS/2005.29). World Health Organization. https://apps.who.int/iris/handle/10665/68985

World Health Organization. (2015, May). World Health Organization Best Practices for the Naming of New Human Infectious Diseases. https://apps.who.int/iris/bitstream/handle/10665/163636/WHO_HSE_FOS_15.1_eng.pdf?sequence=1. Accessed 15 April 2020

World Health Organization. (2018). *A checklist for pandemic influenza risk and impact management 2018 Update*. https://apps.who.int/iris/bitstream/handle/10665/259884/9789241513623-eng.pdf;jsessionid=163B9714EEB4B37BE78B9A8C221C3BF1?sequence=1. Accessed 25 March 2020

World Health Organization. (2020, March 31). *Advice for public*. WHO. https://www.who.int/emergencies/diseases/novel-coronavirus-2019/advice-for-public. Accessed 15 April 2020

Chapter 2
Psychiatric Kinds and the DSM: Notes from a Conceptual Building Site

Rachel Cooper

Abstract The *Diagnostic and Statistical Manual of Mental Disorders* (DSM) is a classification of mental disorders. It is published by the American Psychiatric Association and revised by committee every 15 years or so. This chapter considers the DSM as a 'conceptual building site'. I suggest that in the same way that structural engineers visit building sites to develop their understanding of the possibilities and limitations of designs and materials, conceptual engineers too can benefit from looking at the sites in which concepts are created and refined. Looking at real-life processes of concept revision can enrich our understanding of what the conceptual engineer can achieve and where conceptual engineering might be difficult.

Keywords Conceptual engineering · Diagnostic and statistical manual of mental disorders (DSM) · Psychiatry · Definition of disorder

The conceptual engineer aims to make the world a better place by improving our concepts. They aim to iron-out conceptual muddles, eliminate those concepts that are beyond repair, and to design better new concepts, and so to promote scientific progress, social justice, or some other good.

As with physical engineering, the process of conceptual engineering can be thought of as requiring various steps:

(i) Problem Specification: Before starting work, the engineer must map current problems. They must review our concepts and figure out how they might be defective.
(ii) Design: The engineer must work out a plan that specifies how our concepts might be improved.

R. Cooper (✉)
Lancaster University, Lancaster, UK
e-mail: r.v.cooper@lancaster.ac.uk

M. G. Isaac et al. (eds.), *New Perspectives on Conceptual Engineering - Volume 3*, Synthese Library 515, https://doi.org/10.1007/978-3-031-98535-5_2

(iii) Testing: The engineer should assess whether the planned changes will bring about improvement, through conducting cost-benefit analyses, risk assessment etc. (Isaac et al., 2022).
(iv) Implementation: The engineer must bring about conceptual change.
(v) Maintenance: The new and improved concepts must be robust and stable in use.

To date, much of the literature on conceptual engineering is somewhat abstract and there has been comparatively little discussion of the practicalities of successful conceptual engineering. When authors have anticipated problems, Stage IV: Implementation—ie actually bringing about conceptual change—is where some have expected potential difficulties to arise.

For example, in Herman Cappelen's, 2018 book, *Fixing Language: An Essay on Conceptual Engineering*, Cappelen takes the view that it is worthwhile to attempt conceptual engineering but he thinks it unlikely that projects of conceptual engineering will succeed. This is because Cappelen thinks that changing concepts is extraordinarily hard. In his view, philosophers who propose plans for conceptual revision are in a position somewhat analogous to those who aim at world peace or ideal socialism—only the naïve would expect to achieve their goals. Cappelen's stance leads to a form of noble pessimism

> On the view I defend, the tools we think with are often defective, but there's very little we can do about it... How can this book be a defense of conceptual engineering? . . . I sympathize with this concern, but I also think that it is deeply misguided. It fails to take into account that what I've just described is an almost universal aspect of large-scale normative reflections. Anyone who spends time thinking and talking about large-scale normative matters should do so without holding out too much hope that their talking and thinking will have significant or predictable effects on the relevant aspect of the world (2018, p. 200)

Similarly, in his 2021 paper 'Conceptual engineering is extremely unlikely to work. So what?', James Andow thinks that the odds of any particular project of conceptual engineering succeeding are slim. He urges 'optimism about conceptual engineering on the basis that it seems the prospective pay-offs might more than make up for the small chances of success.' (2021, p. 224).

In this chapter, I am in certain respects more optimistic than Cappelen and Andow, and in certain respects more pessimistic. I suggest that some of the debates around conceptual engineering have become overly abstract. In the same way that structural engineers visit building sites to develop their understanding of the possibilities and limitations of designs and materials, conceptual engineers too can benefit from looking at the sites in which concepts are created and refined.

I seek to enrich our understanding of the potential difficulties that face the conceptual engineer by drawing on examples from a real-life 'conceptual building site'. Specifically, I will draw on examples from psychiatry, where committees regularly revise disorder concepts through revising the sets of diagnostic criteria that define disorders. Looking at real-life processes of concept revision can enrich our understanding of what the conceptual engineer can achieve and where conceptual engineering might be difficult. Drawing on real examples, I will suggest that philosophers should rethink their assumptions about the processes of conceptual engineering. There are grounds for both optimism and pessimism. Some stages of the process of conceptual engineering—specifically, Stage IV: Implementation—Bringing about conceptual change—might in certain contexts be much easier than some scholars have feared. Conversely, looking at real life examples suggests that problems may arise at other stages of conceptual engineering, which in the abstract have been assumed easy.

2.1 The DSM: A Conceptual Building Site

The *Diagnostic and Statistical Manual of Mental Disorders* (more commonly known simply as the DSM), is a major classification of mental disorders published by the American Psychiatric Association (APA, 2013). The contents of the DSM define the domain of psychiatry: mental disorders are those conditions included in the book. For each recognised disorder, the DSM provides a set of diagnostic criteria. For example, for 300.4 Persistent Depressive Disorder, symptoms can include problems with appetite and sleep, fatigue, low self-esteem, poor concentration, and feelings of hopelessness (APA, 2013, pp. 168–169). Characteristically, the symptom lists often require only that a patient display some subset of a longer list (three from eight, say), and specify the length of time for which symptoms must have been present (e.g. a week, 3 months). There are often also various exclusion criteria (e.g. the symptoms must not be caused by some physical condition, or the diagnosis can only be made in patients over age 18). The set of diagnostic criteria thus provides a check list for diagnosis; when a patient displays the requisite symptoms, the diagnosis can be given.

Although published by the American Psychiatric Association, the DSM is influential worldwide. It's true that the World Health Organisation also publishes its own classification of disorders, the *International Classification of Diseases* (ICD), but in recent decades the mental disorders section of the WHO classification has followed the lead of the DSM. Mental health systems are structured by DSM categories; textbooks that are used to train mental health professionals have chapters on DSM categories, psychiatric drugs list the DSM diagnoses for which they are approved, DSM codes are used in hospital admission data and on insurance forms. By such means, the DSM ends up defining which conditions mental health professionals around the world regard as disorders, and how those conditions will be diagnosed. The DSM thus defines psychiatric concepts. Whether a patient will be diagnosed with a mental disorder depends on whether they fit the diagnostic criteria listed in the DSM.

When someone is diagnosed with a mental disorder this can have a huge impact on their life. Most obviously, whether a condition is included or excluded from the DSM affects whether patients will be able to be treated for a condition. Drugs and treatment pathways are developed for DSM diagnoses, and often the bureaucratic systems of health care mean that care can only be provided if some DSM diagnosis or other is given (this is most obviously the case in systems where DSM codes are required by insurance providers). Being diagnosed with a psychiatric disorder also impacts how people are treated by others and how they come to think of themselves. Many mental health conditions are heavily stigmatised; neighbourhoods may whisper about diagnoses of schizophrenia or personality disorders, for example. The effects that psychiatric concepts have on the lives of diagnosed people are considerable—and they are not all positive. Those who are diagnosed with a psychiatric label may be able to access therapies, which they may find helpful, but they may also suffer drug side effects and are likely to suffer various forms of stigma.

As the DSM is of considerable practical significance, the classification is ever contested; groups constantly lobby the American Psychiatric Association with the aim of having this or that condition included or excluded from the classification. Over time new research also accumulates, which may suggest that diagnostic criteria should be revised in some way or other. For example, research may suggest that two conditions with separate listings are importantly similar and so might be best grouped together, or, conversely, that conditions lumped together might be better split. As a result of such pressures, the DSM has to be regularly revised, and the American Psychiatric Association produces a new edition every 15 years or so.

Revising the DSM is a major undertaking. Producing the latest substantive revision, DSM-5 (2013), took many hundreds of committee members many years. In this chapter, I focus on the processes of constructing and revising the DSM,[1] as I suggest that the committees that revise the DSM can reasonably be understood as being engaged in projects of conceptual engineering. In deciding which conditions should be counted mental disorders, and in revising sets of diagnostic criteria, these committees construct and revise psychiatric concepts.

Some might question whether it is fair to count the revision of diagnostic criteria as involving the revision of concepts. Might we instead think of DSM revision as merely being a process whereby descriptions of natural kinds of mental disorder are refined in the light of new empirical evidence, rather than as a process that involves conceptual engineering? It's true that the committees that revise the DSM review huge quantities of empirical data, but much of their decision making is at least in part definitional or conceptual, and for this reason I think it fair to describe such committees as engaging in conceptual engineering. Empirical evidence radically underdetermines the classification of mental disorders for a number of reasons: (i) many mental disorders shade into normality, and also into each other, and so decisions are required in threshold-setting (for example, in DSM-5 the number of symptoms required for diagnosing Attention-Deficit/ Hyperactivity Disorder in adults was reduced), (ii) the causal structure of psychopathology is complex, leaving room for decisions regarding lumping versus splitting (for example, in DSM-5 Hoarding was upgraded from being a possible manifestation of Obsessive Compulsive Personality Disorder to being a distinct disorder), (iii) conceptual

[1] I describe the process of revising the DSM and debates around the most recent edition, the DSM-5 (2013) in greater detail in Cooper (2018).

questions regularly arise as to whether some condition should count as a disorder (for example, in DSM-5 grief can be diagnosed as depression, whereas previously moderate depressive symptoms in bereaved people were considered 'normal' and so not diagnosed as mental disorder) (APA, 1994, 2013).[2]

As the DSM committees decide how to revise the DSM, they pursue the fundamental objectives of conceptual engineering; they aim to make the world a better place by engineering better concepts. Most obviously, DSM committees seek to engineer psychiatric concepts that will be good for science. In specifying diagnostic criteria they aim at 'validity' (i.e. accurately mapping the causal structure of the domain of psychopathology) and 'reliability' (i.e. specifying criteria such that the same patient will be given the same diagnosis by different clinicians). DSM committees also seek to engineer the DSM to benefit patients. Guidelines for those proposing new diagnoses for the DSM-5 asked for a consideration of whether "... the harm that arises from the adoption of the proposed diagnosis exceed[s] the benefit that would accrue to affected individuals" (Kendler et al., 2009, p. 6). Of course, in some cases, these various objectives of 'validity', 'reliability', and 'avoiding harm to patients' can conflict, and decisions then have to be made which to prioritise.

The processes of DSM revision offer an excellent case study for thinking through the potential and difficulties of conceptual engineering, as the committees that shape the DSM have been engaged in conceptual engineering repeatedly and for many years. The DSM further serves as good case study as the processes of revising the DSM are comparatively transparent. Throughout the revision process, the DSM committees publish review papers and commentaries that document much of their thinking, and the American Psychiatric Association maintains archives of DSM committee minutes that I have been able to consult. Through looking at the construction of DSM categories, we can thus get to see examples of conceptual engineering in practice.

At the outset it is worth noting that psychiatric concepts have much in common with other 'human kinds', as described by Ian Hacking (1986, 1992, 1995). Human kinds are kinds of people, as studied by the human sciences. Other examples include race and gender kinds. Hacking notes that human kinds typically come to be employed in structuring bureaucracies. How people are classified—whether they receive this or that diagnosis, whether they are considered a woman or man—makes a difference to what people are permitted do, and what is done by others to them. As human kind classifications are of practical significance and morally-loaded, the classifications are typically contested. Individuals and groups will lobby to be considered 'disordered', or not, or to counted as 'women', or not. Such activities are sometimes successful, and human kind classifications are characteristically unstable.

[2] Relevant pages are ADHD DSM-IV (APA, 1994, p. 83) DSM-5 (APA, 2013, p. 59); Hoarding DSM-IV (APA, 1994, p. 673), DSM-5 (APA, 2013, p. 247). Depression and grief DSM-IV (APA, 1994, p. 327), DSM-5 (APA, 2013, p. 161).

Generally, human kind concepts have mixed effects on people's lives—they make lives go better in some ways, and worse in others. Given that human kind concepts almost always cause at least some people some harm, as a class these concepts are attractive targets for conceptual engineering. Human kind concepts look to be the types of concepts that might often be suboptimal and that might be improved.

However, though there are many similarities between psychiatric concepts and other human kinds we should note key ways in which psychiatric concepts and some other human kinds differ. While psychiatric kinds and some other human kinds are 'committee human kinds', there are also 'folk human kinds'. 'Committee kinds' are those where some committee has authority over such concepts, and has the ability to intentionally revise them. 'Committee kinds' often have transparent origins; a scientific or bureaucratic committee can intentionally construct them (often in minuted meetings). In contrast, in the case of 'folk human kinds' (such as 'child', or 'woman') in most contexts no recognised group of experts has the authority to bring about revisions, and the origins of 'folk human kinds' are generally obscure and hidden in the past. The distinction between 'committee kinds' and 'folk kinds' is not a sharp distinction, but rather a matter of degree and sometimes of context. There may be middling cases—where committees have limited power to shape concepts (they can make suggestions but not ensure uptake), or have authority over some concept but only within particular contexts. When seeking to generalise from the case of the DSM, we can expect there to be many similarities between psychiatric concepts and other human kind concepts that fall within the remit of some scientific or bureaucratic committee, but should expect to find some important differences between psychiatric concepts and 'folk' human kind concepts.

In this chapter, I start by discussing a number of examples from the DSM 'conceptual building site' that look promising for projects of conceptual engineering—these are examples where committee members spot a sub-optimal concept, that is causing some problems, intentionally revise it, and bring about improvement (facilitating either better science or improvements for patients). I then move on to consider some examples that are more puzzling or problematic. The second part of the chapter considers a number of cases that suggest some of the pitfalls that can befall projects of conceptual engineering.

2.2 The Good News: Examples that Illustrate the Potential of Conceptual Engineering

2.2.1 DSM Committees Succeed in Revising Concepts

The first and most obvious way in which the processes of revising the DSM offers encouragement for would-be conceptual engineers is that here we have a case where committees regularly and intentionally succeed in revising concepts. In some writings on conceptual engineering (most prominently in the works of Cappelen, previously cited) it has been suggested that implementing conceptual change might be very difficult. Cappelen is committed to metasemantic externalism that implies that the processes that underlie conceptual change will generally be obscure and uncontrollable. However, the DSM provides a case that suggests that in certain domains bringing about controlled conceptual change might not be so difficult.

In outline, the process by which DSM committees regularly succeed in bringing about conceptual change in a controlled manner is as follows: The DSM committees alter the conditions listed in the DSM as mental disorders, or change the diagnostic criteria that define them. This on its own may not be enough to bring about conceptual change. It is likely that many users of the DSM do not read it closely, and if revisions were simply inserted into the text there is a risk they might pass unnoticed. To make sure that revisions produce conceptual change, when the DSM is revised, the American Psychiatric Association works to ensure that revisions are noticed and followed by mental health professionals. Press releases will highlight major changes. The textbooks published by the American Psychiatric Association are revised, and training courses will guide mental health clinicians through the new manual.

Such means ensure that when the DSM is revised, mental health professionals know about the revisions. Disciplinary, professional and bureaucratic pressures then act to ensure that mental health clinicians and researchers will come to employ the new concepts that the DSM committees have constructed. Researchers seeking grants or publishing papers will be expected by referees to talk the new DSM-talk. The forms used by health care bureaucracies to organise patient care and payment will similarly force clinicians to think in DSM terms.

When the DSM committees revise concepts, there are mechanisms that ensure that communities of mental health professionals will come to use these new concepts. As such, the DSM committee members are in a position to intentionally bring about conceptual change. The power that the DSM committees have to bring about conceptual change is somewhat unusual, but not unique. As well as in psychiatry, there are many other fields where committees of scientists or bureaucrats have the power to bring about conceptual change. Roughly, what is required is for some body of experts to have the authority to ensure that some community will change their linguistic and classificatory practices in accord with the conceptual revisions they propose. There are many fields of science where committees responsible for nomenclature and classification have this power. For example, the International Committee on Taxonomy of Viruses rules on the nomenclature and classification of viruses (Adams et al., 2017), while the International Astronomical Union rules on astronomical terms (famously, redefining 'planet' in 2006) (IAU, 2006). Governmental and bureaucratic bodies also often have similar authority to construct concepts and to ensure they are employed in at least those contexts under their control—for example the UK Equality Act 2010 defined 'disability' so as to include long term illness, and as such a definition is employed in benefits payments and employment law, at least partial uptake is assured (Office for Disability Issues, 2011).

2.2.2 *An Example of Ameliorative Conceptual Engineering in the DSM: The Definition of Disorder*

In addition to extensionally defining 'mental disorder' through its contents list, since 1980 the DSM has included a definition of mental disorder in its introduction (APA, 1980, p. 6). This definition evolved during debates in the late 60s and early 70s over homosexuality. The DSM-I and DSM-II listed homosexuality as a mental disorder (APA, 1952, 1968). At that time, thinking of homosexuality as a disorder was not peculiar to the DSM but commonplace amongst psychiatrists internationally. During the 1960s and 1970s, classifying homosexuality as a mental disorder became increasingly controversial. Gay activists began to lobby the APA to get homosexuality removed from the DSM (their efforts and the development of the debates are detailed in Bayer, 1981). Protests came to a head in 1970 when gay rights protesters targeted the American Psychiatric Association annual meeting in San Francisco, shouting down protesters with whom they disagreed and disrupting much of the meeting.

During the late 1960s and early 1970s, whether homosexuality should be considered a disorder became a hotly debated issue. Senior psychiatrists argued both sides of the debate; some claiming that homosexuality was a pathological deviation, others that homosexuality should be considered a normal variant of human sexuality. Eventually, the American Psychiatric Association resolved the issue through a vote of its members, and in 1973 homosexuality was removed from the DSM.

By the time of the vote, a consensus had developed that homosexuality should not be considered a disorder. Both empirical evidence and conceptual arguments played a role in shaping this consensus. At the empirical level, a number of studies suggested that homosexuality is more common and that homosexual people lead happier lives than previously expected. But conceptual engineering also played a major role. One strand of the debates over homosexuality focussed on the question of how 'mental disorder' should be defined.

Robert Spitzer became a key figure in these discussions. Spitzer's position was that homosexuality in itself is not a disorder (Spitzer 1973, 1981). To defend his claim, Spitzer developed a definition of 'mental disorder'. According to Spitzer's definition, a condition can only be a mental disorder if it causes distress or disability. The best developed defence of his position can be found in a 1978 paper written jointly with Jean Endicott, which defines mental disorders as follows:

> A relatively distinct condition resulting from an organismic dysfunction which in its fully developed or extreme form is directly and intrinsically associated with distress, disability, or certain other types of disadvantage. (Spitzer & Endicott, 1978, p. 18)

Spitzer claimed that all the disorders included in the DSM-II, with the exception of homosexuality, caused harm (distress or disability or risk thereof). Given that many homosexuals are perfectly happy being gay, on Spitzer's account homosexuality in and of itself cannot be a disorder.

Spitzer and Endicott's paper looks recognisably to be a project in conceptual analysis. In so far as their analysis contributed to shaping notions of mental disorder amongst mental health professionals rather than simply describing a pre-existing concept, I think their efforts can reasonably be regarded as an instance of ameliorative conceptual engineering. In their paper Spitzer and Endicott propose a definition of mental disorder in terms of necessary and sufficient conditions. Once they have got a definition that they think generally works, they then employ it to defend their position that homosexuality should not be considered a disorder.

Following the debates over homosexuality, Spitzer became a powerful figure in US psychiatry, when he was appointed chairman of the committee responsible for developing the DSM-III (1980). A consensus developed amongst mental health professionals that harm is necessary for disorder and a version of Spitzer's definition of mental disorder came to be included in the introduction of the manual.

Here we have what looks to be a straightforward example of successful ameliorative conceptual engineering, which involved engineering a new concept of mental disorder whereby harm was necessary for disorder: Homosexuality used to be considered a mental disorder; this caused problems for gay people. The definition of mental disorder was clarified and revised. As a result, it was made clear that a condition can only be a disorder if it causes harm, and homosexuality was subsequently removed from the list of mental disorders.

2.3 The Bad News: Examples that Illustrate the Difficulties of Conceptual Engineering

The examples discussed above show how conceptual engineering may sometimes be possible. Other examples from the DSM, however, illustrate potential difficulties that may face the conceptual engineer.

2.3.1 The Consequences of Conceptual Changes Are Hard to Predict

DSM committees try to 'do good' when making revisions, but conceptual change can have unintended harmful consequences, which means that conceptual engineers who aim at improving our concepts can accidentally make things worse (see also Cappelen, 2018, p. 159; Marques, 2020). The committees in charge of deciding which diagnoses to add to the DSM, and how to shape the diagnostic criteria for those diagnoses that are included, try to anticipate the effects of revision. Review papers sketch the hoped for benefits and potential problems of proposed changes (for example Boelen & Prigerson, 2012; Huprich, 2012; Mataix-Cols et al., 2010; Selby et al., 2012; Woods et al., 2010). Very often, however, such predictions of costs and benefits turn out to be radically wrong. Anticipating the consequences of changing diagnostic criteria turns out to be exceptionally difficult. New psychiatric concepts simultaneously create new concepts for scientific research and new bureaucratic concepts. How people respond to diagnostic labels, what scientific research might discover, and how bureaucracies will come to treat people are all hard to predict. These various sources of unpredictability can also interact, creating a 'triple-whammy' effect, and making it hard to anticipate the consequences of revising psychiatric concepts.

2.3.1.1 Difficulties of Predicting How People Will Respond to Diagnoses

As Ian Hacking has emphasized in his work, it is very hard to predict how the creation of new human kinds will affect people (Hacking, 1986, 1992, 1995). When human kind labels are used to describe people, those who are labelled are not passive recipients. They respond in various ways; labels may be resisted or embraced, and the narratives that are told about kinds of people develop over time.

Take for example the connotations of being diagnosed as having 'autism' or 'Asperger's disorder'. When autism was included in the DSM-III (1980) it was thought of as being a severe and very rare condition. According to DSM-III, autism is 'extremely incapacitating, and special educational facilities are almost always necessary' (APA, 1980 p. 88). Prevalence was estimated to be 2–4 cases per 10,000 (APA, 1980, p. 89). In 1980, the condition was heavily stigmatized—though those receiving the diagnosis were then thought incapable of understanding its meaning. In 1994, Asperger's Disorder was added as a distinct diagnosis in the DSM-IV. Asperger's was then thought of as being a condition a little less severe than classical autism and was also expected to be rare (Volkmar et al., 1994). In reality, however, after the publication of DSM-IV, there was a massive increase in autism-related diagnoses (Pinborough-Zimmerman et al., 2012), and, in particular, many people came to be diagnosed with Asperger's Disorder. Over time, Asperger's Disorder became relatively destigmatized; communities developed around the Asperger's label, and many people came to self-identify as 'Aspies'.

Our story continues: With the publication of DSM-5 in 2013, Asperger's was removed as a standalone diagnosis from the DSM. It was argued that the distinction between Asperger's and other autism-related conditions could not be reliably drawn (Happé, 2011). Instead the DSM-5 included a new category—Autistic Spectrum Disorder (ASD)—which was intended to cover both those previously diagnosed with autism as well as those previously diagnosed with Asperger's disorder. In the lead up to the DSM-5 being published, there was substantial debate as to the potential consequences of the removal of Asperger's as a distinct condition from the DSM. Many feared that getting rid of Asperger's Disorder would cause harm to communities that identified as 'Aspies' (e.g. Kite et al., 2013, Giles, 2014). However, in the event, initial evidence suggests that little such harm occurred. Some of those who previously identified with 'Asperger's' simply came to re-identify as 'autistic' (a label which is also becoming destigmatised) (Petrovic & Petrovic,

2021), while others who previously identified as 'Asperger's' simply carried on using the label despite its removal from the DSM (e.g. Asperger/ Autism Network (AANE), 2022).[3]

In this history, the responses of diagnosed people turned out to be radically unpredictable. In 1994, it was thought that diagnoses of Asperger's would be rare and stigmatised. They turned out to be common and relatively socially acceptable. In 2013, it was feared that the loss of the 'Asperger's' label might harm those who self-identified as 'Aspies', but it looks like this has turned out not to be the case. As this, and Hacking's case studies show, it is generally hard to predict how people will come to respond to being labelled with human kind terms.

2.3.1.2 Difficulties of Predicting the Effects of New Concepts for Scientific Research

New psychiatric concepts create new concepts for scientific research. When a new diagnosis comes to be included in the DSM, it is often hoped that this will enable research into the condition, which may eventually result in some efficacious therapy being developed. However, whether such research will succeed will be unknown. Even if a drug for a condition becomes available, whether its use on balance will be a good thing may be uncertain. Treatment with psychiatric drugs often, but not always, benefits patients. Side effects are particularly a risk in psychiatry as many psychiatric drugs are taken over long periods of time.

When the DSM-5 was under development, there were proposals for the inclusion of a new disorder, Attenuated Psychosis Syndrome. This would have been a diagnosis for young people considered to be at risk of developing schizophrenia. In the event, this diagnosis was not included in the DSM-5. A major concern was that the diagnosis might lead to many odd teenagers being prescribed antipsychotics and suffering the side effects of these drugs for little benefit (Moran, 2009).

Whether Attenuated Psychosis Syndrome would actually have been a dangerous diagnosis is hard to say, and would have depended on what future scientific research discovered. While antipsychotic drugs have nasty side-effects, some studies have suggested that Attenuated Psychosis Syndrome might be also be effectively treated with fish oils (Amminger et al., 2010). If these findings were confirmed, or some other harmless treatment was discovered, then the context of the debate would shift radically. Generally, the outcomes of future research are hard to foresee.

[3] In general, I hold that DSM committees have the power to reshape psychiatric concepts. However Asperger's is a case where a psychiatric concept became a lay concept and so has escaped their control.

2.3.1.3 Difficulties of Predicting the Effects of New Bureaucratic Kinds

In addition to being concepts used in scientific research, DSM diagnoses also commonly serve bureaucratic functions. DSM diagnoses appear in legislation, and in government and bureaucratic policies. By such means, the diagnosis that someone receives can massively impact their life: children with this or that diagnosis come to be provided with specific educational services, the diagnosis that an offender receives can affect how they are dealt with by the legal system. How a new psychiatric concept will interact with bureaucratic systems, and so what effect it will have on people's lives can be hard to predict. Policies are prone to revision, and new laws or new guidelines can utterly reshape the implications of receiving a particular diagnosis.

In *The Book of Woe* (2013), Gary Greenberg interviews Allen Frances, who was chairperson for the DSM-IV and DSM-IV-TR (1994, 2000). Frances is now haunted by some of the accidental consequences of the revisions he oversaw.

In DSM-IV-TR the diagnostic criteria for many paraphilias (sexual perversions) required that the person had 'acted on these urges or is markedly distressed by them'. In the words of Frances, this turned out to be 'one royal fuckup' (Greenberg, 2013, p. 233). The unintended consequence of the 'or' in the criteria was that those who commit sex crimes could be diagnosed with a paraphilia in virtue of their crime alone. The DSM-IV-TR included a diagnosis 'Paraphilia: Not otherwise specified (NOS)'—this was intended to provide a code for rare paraphilias for which no separate diagnostic criteria were provided, e.g. for the diagnosis of patients whose sexual interests and activities revolve around corpses, or faeces, or vampires. However, beginning in the late 1990s, diagnoses of Paraphilia, NOS (Nonconsent) started to be employed in legal cases to describe some rapists. Once successfully employed in a few cases, legal appeals to Paraphilia NOS: Nonconsent quickly multiplied (King et al., 2014). In the context of the US legal system, a diagnosis of paraphilia took on added significance, as the sexually violent predator laws in many states meant that offenders with a paraphilia diagnosis could be detained indefinitely. None of this could have been predicted by the DSM committees, who did not expect that rapists would come to be diagnosed as mentally disordered.

When new psychiatric concepts are constructed it is very hard to predict (i) how diagnosed people will respond to the new label, (ii) what scientific research will discover about the new diagnoses, (iii) how legal systems and bureaucracies will make use of the diagnosis. DSM committees seek to 'do good' when they revise the classification, but predicting which revisions will on balance benefit patients is very hard. Similar problems can be expected to face other would-be conceptual engineers. Many concepts are at one and the same time (i) labels that are applied to people (ii) concepts for scientific research, (iii) concepts employed in legal and bureaucratic contexts.

2.3.2 *Conceptual Muddles Can Be Hard to Spot, and Not all Linguistic Communities Value Clarity*

It is a fundamental working assumption of much work in the philosophy of science that conceptual confusions are liable to cause problems and should generally be fixed. An extremely common structure for papers in the philosophy of science is for a philosopher to claim to uncover some hidden ambiguity in a scientific concept, and then to assume that clarifying this concept is necessary for enabling scientific progress (as illustrative examples see Millikan. 1989 on 'function', Sakar, 2002 on 'biodiversity', Laplane 2016 on 'stemness'). In so far as those identifying conceptual confusions depend on a reading of scientific texts, it is further assumed that concepts employed by a scientific community can be mapped through reading scientific texts. Some examples from the DSM challenge the assumptions: it may be harder than many have assumed to identify conceptual muddles through reading scientific texts, and some scientific communities may be surprisingly tolerant of conceptual confusion.

2.3.2.1 Slippage Between Text and Practice: Problems for Those Seeking to Map Conceptual Muddles

The DSM-IV (1994) contained a diagnosis called 'Pervasive Developmental Disorders Not Otherwise Specified' (PDD-NOS). The intention of this category was to provide a label for autism-like conditions that were atypical in some respect or other (e.g. with less severe symptoms, or occurring at a later age of onset). The DSM-IV defined PDD-NOS as follows

> This category should be used when there is a severe and pervasive impairment in the development of reciprocal social interaction or verbal and nonverbal communication skills, OR when stereotyped behavior, interests, and activities are present, but the criteria are not met for a specific Pervasive Developmental Disorder... (APA, 1994, pp. 77–78 [emphasis added])

In this definition, the 'or' highlighted should have been an 'and'. As written, the definition would allow for the diagnosis of someone who simply manifested 'stereotyped behavior, interests, and activities' (e.g. someone who devoted their life to playing scrabble, or to understanding modal logic), but who had no other difficulties. This was not at all what the DSM committees intended.

In the event this error in the DSM-IV seems to have caused no problems. In the contexts in which PDD-NOS is diagnosed, it seems few clinicians noticed the mistake. Rather than following the literal text, it seems these clinicians just had a general picture of what PDD-NOS might look like, taking it for granted that some autism-like condition was required, and diagnosed accordingly.

The mistake in the criteria for PDD-NOS was corrected in the next edition, DSM-IV-TR—but it's likely that few readers noticed the correction. Usually readers are alerted to revisions to DSM diagnostic criteria but the correction to PDD-

NOS occurred quietly. By the time the DSM-IV-TR was under development, the American Psychiatric Association was battling accusations that it revised the DSM too frequently. Amongst such criticism, the DSM-IV-TR was marketed as being a new edition that contained no substantial revisions to sets of diagnostic criteria. As well as the sets of diagnostic criteria, the DSM includes many pages of descriptive text, which describe what is known about the etiology of each disorder, its natural history, cultural variations, and so on. DSM-IV-TR was sold as a new edition in which only this descriptive text had supposedly been revised (the 'TR' in DSM-IV-TR, stands for 'Text Revision'). In actual fact, although the marketing blurb suggested that the sets of diagnostic criteria in the DSM-IV-TR were the same as in the DSM-IV, this was not quite true. There were a few changes made to sets of diagnostic criteria—the correction to the criteria for PDD-NOS amongst them—but as readers were not alerted to these changes, it's likely that few noticed them.

Here we have a case where the diagnostic criteria for PDD-NOS contained a mistake in DSM-IV, and this mistake was corrected in DSM-IV-TR. But it's likely that few readers ever noticed the original error, and, as the correction was made quietly and not advertised, it's also likely that few readers noticed the correction.

One might say, so what? This example just shows that psychiatrists don't always read the DSM very well. I think the case troubling. It illustrates how there can be slippage between written-text and read-text such that it's unclear what methodology might be appropriate for a conceptual engineer who is seeking to survey the concepts employed in a field. Traditionally, philosophers read texts carefully and literally and look for conceptual errors (contradictions, ambiguities, odd counter-examples and so on). But here we have a case where text and practice come apart in ways that make it unclear whether the concepts of a field can be extracted from texts. Rather, this case suggests that sometimes philosophers seeking to survey the conceptual adequacy of a field will need to look elsewhere—at spoken language use and classificatory practices. Philosophers of science have tended to assume that surveying the conceptual adequacy of some domain will be straightforward (they read the texts and looks for problems). The above example shows that more sophisticated, and more empirically grounded, methodologies might also be needed.

2.3.2.2 A Tricky Case: A Deliberately Inserted Contradiction

Philosophers of science have tended to take it for granted that conceptual muddles (most obviously contradictions) are bad, and have assumed that all right-thinking researchers will be motivated to avoid them. Some examples from the DSM problematise these foundational assumptions.

As discussed previously, the DSM includes a definition of mental disorder in its introduction. The DSM-IV (1994) introductory definition stated that a condition can only be considered a disorder if it causes harm (p. xxi). In the DSM-IV a requirement that symptoms cause harm was also added to many of the sets of diagnostic criteria (usually worded 'causes clinically significant distress or impairment in social, occupational, or other important areas of functioning') (APA, 1994, p. 7). The DSM-

IV committee took it that the general introductory definition of disorder already stated that disorders cause harm and that adding a requirement that there be distress or impairment to sets of diagnostic criteria merely emphasised the point (First et al., 1995).

Tics are involuntary movements or vocalisations, and tic disorders are conditions like Tourette's. According to the DSM-IV diagnostic criteria for tic disorders, such disorders can only be diagnosed if the tics cause 'marked distress or significant impairment in social, occupational, or other important areas of functioning' (APA, 1994, pp. 103–105). However, on publication of the DSM-IV, it turned out that some researchers specialising in tic disorders disagreed that 'distress or impairment' should be a requirement (Freeman et al., 1995; Comings, 1995).

The issue is that many people tic without their tics causing them any problems. In particular, many young children who tic do not care at all. According to the DSM-IV, although people who tic but where their tics cause no problems can fairly be described as having tics, they should not be diagnosed as having a tic *disorder*, as a disorder can only be diagnosed when there is harm. Some researchers who specialise in tic conditions consider this wrong-headed. They think that all people who tic should be considered to have tic disorders, regardless of whether an individual happens to experience problems because of their tics or not.

In the next edition of the DSM, DSM-IV-TR (2000), in the introduction, the general definition of mental disorder remains unchanged and, as before, this states that a disorder can only be diagnosed if there is harm (APA, 2000, p. xxxi). However, in response to the concerns of tic researchers, the criterion that symptoms cause 'distress or impairment' was removed from sets of diagnostic criteria for the tic disorders.

It is hard to know what to make of these revisions. On the assumption that the introductory definition of mental disorder applies to all disorders in the classification, removing the 'distress or impairment' criterion from tic disorders should make no difference to who is diagnosed; when particular sets of diagnostic criteria include 'distress or impairment' this merely emphasises a point that is already implicit. On the other hand, in so far the tic researchers were appeased by the revision, it seems that it must have been expected to make some difference in practice.

Plausibly, the explanation for the fact that the revisions led to an inconsistency between the diagnostic criteria for tic disorders and the introductory definition of disorder, and that this inconsistency seemed tolerable to both critics and those revising the DSM, is that it is tacitly accepted that no one is expected to read the whole manual. Different groups contribute to, and read, different portions of the text; so long as 'their' sections are as they wish, they don't always care if the DSM contains contradictory statements elsewhere. In the DSM-IV-TR, any researchers who want to diagnose harmless tics, can just look at the tic disorder section. At the same time, anyone who wants to diagnose disorders only when there is harm, can appeal to the introduction.

There are two ways of reading this case. One might conclude that a conceptual engineer seeking to map DSM-concepts should read it piecemeal rather than as a whole. The concept of disorder implicit in the tic disorders section is simply different from that elsewhere, and properly interpreted there is context-relativity rather than contradiction. Go down this route, however, and one must abandon the notion that mapping the conceptual health of a domain will be easy (how should the DSM properly be split? by chapter? or on some sociological basis?). Alternatively, if one takes the DSM as a whole, then one must conclude that psychiatry is prepared to be tolerant of contradictions. Take this route, and this problematises the notion that all serious, scientific, professional fields, will prioritise the avoidance of conceptual muddles.

2.3.2.3 Another Case: The Definition of Disorder—Act 2

As discussed in Sect. 2.2, following the debates over the status of homosexuality, the DSM came to include an introductory definition of mental disorder that stated that a condition could only be considered a disorder if it caused harm. However, in the DSM-5 (2013), the definition of mental disorder was changed, and the story of the changes to the definition poses a number of challenges for thinking about conceptual engineering,

Although the DSM-5 was published in 2013, work on the edition started long before. By late 2011 the revisionary process was well advanced and a close to final draft of the DSM-5 was made available for feedback. By this stage, the revisions to the diagnostic criteria for the various diagnoses included in the DSM, had all fairly much been agreed. However, one aspect of the DSM-5 that was still up for discussion was the introductory definition of disorder.

In the 2011 draft, two competing definitions of mental disorder, developed by different groups, were presented, and a note explained that a decision would be made later which to adopt.[4] The first definition was a minimal revision of the definition included in the DSM-IV-TR. This definition was developed by a group which included some philosophers and continued to state that harm is necessary for a condition to be a disorder (Stein et al., 2010). The second, competing definition was developed by the Impairment and Disability Assessment Study Group, chaired by Jane Paulsen, which was charged with reviewing the links between the diagnosis of disorder and notions of distress and disability in the DSM. In this definition all reference to distress and impairment had been removed.[5]

[4] On publication of the DSM-5, the American Psychiatric Association deleted the draft versions. However, a draft that was available December 2011, and that has the two definitions, can still be accessed at https://web.archive.org/web/20111209002319/http://www.dsm5.org/ProposedRevisions/Pages/proposedrevision.aspx?rid=465.

[5] In outline, the motivation of this group was to bring the DSM definition of disorder closer to that implicit in the International Classification of Diseases (ICD), published by the World Health Organization (WHO), which has been influenced by the distinction between 'impairment' and 'disability' employed by advocates of the Social Model of Disability. I discuss the details further in Cooper (2015).

Within psychiatry, the proposed changes to the definition of disorder were little discussed. No one seemed to care much how 'mental disorder' was defined. Eventually, when the DSM-5 was published, the definition of mental disorder included in the introduction read as follows:

> A mental disorder... reflects a dysfunction in the psychological, biological, or developmental processes underlying mental functioning. Mental disorders are USUALLY ASSOCIATED with significant distress or disability in social, occupational, or other important activities. (APA, 2013, p. 20, emphasis added).

What happened? As discussed in Sect. 2.2, from the late 1970s to 2012, it looked like those DSM committees that drafted the definition of disorder for the DSM were engaged in a project that looked recognisably like an attempt define mental disorder in terms of necessary and sufficient criteria. And then, with the DSM-5, this project was abandoned. We can note that a definition that says that disorders are 'usually associated' with harm seems to be of little use in determining who should be considered to have a mental disorder. The definition gives no guidance as to whether any particular 'symptomatic' but flourishing individual should be diagnosed. It is also unclear whether it's types or tokens of mental disorders that are 'usually associated' with distress or disability (ie is it that most types of mental disorder are associated with harm, or that most people with any particular mental disorder will experience harm).

The published DSM-5 definition appears to be an uncomfortable compromise between the definition developed by Stein et al. and by the Impairment and Disability Assessment Study Group. The final definition was produced late on in the process of revising the DSM, and there is no publicly accessible explanation of its construction, but it looks to be a compromise that makes no real sense—a mix-and-match mash-up drafted by some committee that has little respect for the niceties of necessary and sufficient conditions.

However, although the new definition is ambiguous, it is important to note that this bothers few mental health researchers or clinicians. During the 1970s debates around homosexuality, those responsible for the DSM found it useful to define 'mental disorder' in terms of necessary and sufficient conditions. Now, however, DSM committees have no appetite for such activities. In this example, we see that not all communities will value clear definitions, and that a community that was once interested in conceptual precision can lose interest. Some have assumed that once a conceptual engineer has completed their job, and fixed some concept, their work will be done, and that future generations will benefit from their labours. Andow (2021) for example, thinks that bringing about conceptual change is very hard, but that if successful this 'might mean no further such work is necessary for centuries to come' (p. 216). In the example of the changing definitions of disorder in the DSM I think we have a case that suggests that conceptual change can sometimes be fairly easy to bring about, but not necessarily robust. I think that DSM committees have the power to reshape concepts—but that what one committee achieves can be undone by the next.

More generally, the example illustrates that while philosophers of science have tended to take it for granted that conceptual clarity is a good thing, not all scientific communities will concur. Philosophers often assume that scientific communities will seek to avoid contradictions, and be motivated to clarify ambiguities (such claims are perhaps most clearly articulated in LaPorte (2004), though for an opposing view see Édouard Machery this volume). However, we must remember there are contexts in which some judge vagueness, ambiguity, and outright contradiction to be tolerable. Here, in these examples, we see DSM committees using conceptual unclarity as a mechanism for 'compromise'. Similar phenomenon have been described elsewhere. For example, in his work on legal theory, H.L.A. Hart (1961) discusses the usefulness of the 'open texture' of language in law (ie it can be beneficial for laws to allow for a range of interpretations), and some work on the communication strategies employed in business similarly points out the strategic usefulness of being unclear (for example, Eisenberg, 1984; Giroux, 2006). Philosophers of science have often taken it for granted that contradictions and ambiguities should be fixed—but others may need convincing.

2.4 Conclusions

Structural engineers visit building sites to develop their understanding of the possibilities and limitations of building design and materials. In this chapter I have shown that it may also be useful for the conceptual engineer to examine 'conceptual building sites' and so to gain a deeper appreciation of the potential and limitations of conceptual engineering. I have discussed examples from the DSM, though case studies looking at other sites where concepts are built could be similarly useful.

Using examples from psychiatry, I suggest that some aspects of conceptual engineering may be easier than some philosophers have anticipated, while other aspects may be much harder. I have argued that in certain contexts, bringing about conceptual change can be much easier than some philosophers have feared. DSM committees succeed in revising concepts in a fairly controlled manner every time the DSM is revised.

However, there are also some reasons for pessimism. Changing concepts may be fairly easy, but changing concepts in ways that succeed in improving the world is harder, as the as long-term effects of conceptual changes can be hard to predict. I have also discussed examples that problematise some of the basic assumptions of much philosophical conceptual engineering. Philosophers of science have tended to assume that conceptual muddles can be identified by reading texts, but we have seen cases where slippage between texts, interpretation, and practice, problematises spotting and correcting conceptual confusions. Philosophers of science have also often assumed that all right-thinking people will see the desirability of conceptual clarity, but we have seen that not all scientific or professional communities will be interested in philosophy-style conceptual engineering.

References

Adams, M. J., Lefkowitz, E. J., King, A. M. Q., et al. (2017). 50 years of the international committee on taxonomy of viruses: Progress and prospects. *Archives of Virology, 162*, 1441–1446.

American Psychiatric Association. (1952). *Diagnostic and statistical manual: Mental disorders*. American Psychiatric Association.

American Psychiatric Association. (1968). *Diagnostic and statistical manual of mental disorders* (2nd ed.). American Psychiatric Association.

American Psychiatric Association. (1980). *Diagnostic and statistical manual of mental disorders* (3rd ed.). American Psychiatric Association.

American Psychiatric Association. (1994). *Diagnostic and statistical manual of mental disorders* (4th ed.). American Psychiatric Association.

American Psychiatric Association. (2000). *Diagnostic and statistical manual of mental disorders* (4th edition text revision ed.). American Psychiatric Association.

American Psychiatric Association. (2013). *Diagnostic and statistical manual of mental disorders* (5th ed.). American Psychiatric Association.

Amminger, G., Schäfer, M., Papageorgiou, K., Klier, C., Cotton, S., Harrigan, S., et al. (2010). Long-chain ω-3 fatty acids for indicated prevention of psychotic disorders. *Archives of General Psychiatry, 67*(2), 146–154.

Andow, J. (2021). Conceptual engineering is extremely unlikely to work. So what? *Inquiry, 64*(1–2), 212–226.

Asperger/Autism Network (AANE). (2022). *AANE's Use of the Term "Asperger's"*. Retrieved August 18 2022 from https://www.aane.org/aanes-use-of-the-term-aspergers/

Bayer, R. (1981). *Homosexuality and American psychiatry*. Basic Books.

Boelen, P., & Prigerson, H. (2012). Prolonged grief disorder as a new diagnostic category in DSM-5. In M. Stroebe, H. Schut, & J. van den Bout (Eds.), *Complicated grief: Scientific foundations for health care professionals* (pp. 85–98). Routledge.

Cappelen, H. (2018). *Fixing language: An essay on conceptual engineering*. Oxford University Press.

Comings, D. (1995). DSM-IV criteria for Tourette's. *Journal of the American Academy of Child and Adolescent Psychiatry, 34*, 401–402.

Cooper, R. (2015). Must disorders cause harm? The changing stance of the DSM. In S. Demazeux & P. Singy (Eds.), *The DSM-5 in perspective* (pp. 83–96). Springer.

Cooper, R. (2018). *Diagnosing the diagnostic and statistical manual of mental disorders*. Routledge.

Eisenberg, E. M. (1984). Ambiguity as strategy in organizational communication. *Communication Monographs, 51*(3), 227–242.

Freeman, R., Fast, D., & Kent, M. (1995). DSM-IV criteria for Tourette's. *Journal of the American Academy of Child and Adolescent Psychiatry, 34*, 400–401.

First, M., Frances, A., & Pincus, H. (1995). DSM-IV editors reply. *Journal of the American Academy of Child and Adolescent Psychiatry, 34*, 402.

Giles, D. C. (2014). 'DSM-V is taking away our identity': The reaction of the online community to the proposed changes in the diagnosis of Asperger's disorder. *Health, 18*(2), 179–195.

Giroux, H. (2006). 'It was such a handy term': Management fashions and pragmatic ambiguity. *Journal of Management Studies, 43*(6), 1227–1260.

Greenberg, G. (2013). *The book of woe: The DSM and the unmaking of psychiatry*. Blue Rider.

Hacking, I. (1986). Making up people. In T. Heller, M. Sosna, & D. Wellberry (Eds.), *Reconstructing individualism* (pp. 222–236). Stanford University Press.

Hacking, I. (1992). World-making by kind-making: Child abuse for example. In M. Douglas & D. Hull (Eds.), *How classification works* (pp. 180–238). Edinburgh University Press.

Hacking, I. (1995). The looping effects of human kinds. In D. Sperber & A. Premark (Eds.), *Causal cognition* (pp. 351–394). Clarendon Press.

Happé, F. (2011). Criteria, categories, and continua: Autism and related disorders in *DSM-5*. *Journal of the American Academy of Child and Adolescent Psychiatry, 50*, 540–542.
Hart, H. L. A. (1961). *The concept of law*. Oxford University Press.
Huprich, S. K. (2012). Considering the evidence and making the most empirically informed decision about depressive personality disorder in DSM-5. *Personality Disorders: Theory, Research, and Treatment, 3*(4), 470–482.
International Astronomical Union. (2006). Press release. IAU 2006 General Assembly: Result of the IAU Resolution Votes. Press release 24 August 2006. Retrieved August 18 2022 from https://www.iau.org/news/pressreleases/detail/iau0603/
Isaac, M. G., Koch, S., & Nefdt, R. (2022). Conceptual engineering: A road map to practice. *Philosophy Compass, 17*(10), e12879. https://doi.org/10.1111/phc3.12879
Kendler, K., Kupfer, D., Narrow, W., Phillips, K., & Fawcett, J. (2009). Guidelines for making changes to DSM-5. This was originally available at www.dsm5.org/ProgressReports/Documents/Guidelines-for-Making-Changes-to-DSM_1.pdf but has been removed. It can now be accessed via the WayBack Machine (internet archive). Retrieved August 18 2022 from https://web.archive.org/web/20100703020444/http://www.psych.org/MainMenu/Research/DSMIV/DSMV/DSMRevisionActivities/Guidelines-for-Making-Changes-to-DSM_1.aspx
King, C. M., Wylie, L. E., Brank, E. M., & Heilbrun, K. (2014). Disputed paraphilia diagnoses and legal decision making: A case law survey of paraphilia NOS, nonconsent. *Psychology, Public Policy, and Law, 20*(3), 294–308.
Kite, D. M., Gullifer, J., & Tyson, G. A. (2013). Views on the diagnostic labels of autism and Asperger's disorder and the proposed changes in the DSM. *Journal of Autism and Developmental Disorders, 43*(7), 1692–1700.
LaPorte, J. (2004). *Natural kinds and conceptual change*. Cambridge University Press.
Marques, T. (2020). Amelioration vs perversion. In T. Marques & A. Wikforss (Eds.), *Shifting concepts: The philosophy and psychology of conceptual variability* (pp. 260–284). Oxford University Press.
Mataix-Cols, D., Frost, R., Pertusa, A., Clark, L., Saxena, S., Leckman, J., et al. (2010). Hoarding disorder: A new diagnosis for DSM-5? *Depression and Anxiety, 27*(6), 556–572.
Millikan, R. (1989). An ambiguity in the notion of function. *Biology and Philosophy, 4*, 172–176.
Moran, M. (2009). DSM-5 developers weigh adding psychosis risk. *Psychiatric News, 44*(16), 5–36.
Office for Disability Issues, H M Government. (May 2011). Equality Act 2010 Guidance: Guidance on matters to be taken into account in determining questions relating to the definition of disability. Retrieved August 18 2022 from https://assets.publishing.service.gov.uk/government/uploads/system/uploads/attachment_data/file/570382/Equality_Act_2010-disability_definition.pdf
Petrovic, S., & Petrovic, D. (2021). Asperger's or Autism—The right terminology for your child. *Autism Parenting Magazine*, Feb 17 221. Retrieved August 18 2022 from https://www.autismparentingmagazine.com/aspergers-autism-terminology/
Pinborough-Zimmerman, J., Bakian, A., Fombonne, E., Bilder, D., Taylor, J., & McMahon, W. (2012). Changes in the administrative prevalence of autism spectrum disorders contribution of special education and health from 2002–2008. *Journal of Autism and Developmental Disorders, 42*(4), 521–530.
Sarkar, S. (2002). Defining "biodiversity"; assessing biodiversity. *The Monist, 85*(1), 131–155.
Selby, E., Bender, T., Gordon, K., Nock, M., & Joiner, T. (2012). Non-suicidal self-injury (NSSI) disorder: A preliminary study. *Personality Disorders: Theory, Research, and Treatment, 3*(2), 167–175.
Spitzer, R. (1973). A proposal about homosexuality and the APA nomenclature: Homosexuality as an irregular form of sexual behaviour and sexual orientation disturbance as a psychiatric disorder. *American Journal of Psychiatry., 130*, 1214–1216.
Spitzer, R. (1981). The diagnostic status of homosexuality in DSM-III: A reformulation of the issues. *American Journal of Psychiatry., 138*, 210–215.

Spitzer, R., & Endicott, J. (1978). Medical and mental disorder: Proposed definition and criteria. In R. Spitzer & D. Klein (Eds.), *Critical issues in psychiatric diagnosis* (pp. 15–39). Raven.

Stein, D. J., Phillips, K. A., Bolton, D., Fulford, K. W. M., Sadler, J. Z., & Kendler, K. S. (2010). What is a mental/psychiatric disorder? From DSM-IV to DSM-V. *Psychological Medicine, 40*(11), 1759–1765.

Volkmar, F., Klin, A., Siegel, B., Szatmari, P., Lord, C., Campbell, M., et al. (1994). Field trial for autistic disorder in DSM-IV. *American Journal of Psychiatry, 151*(9), 1361–1367.

Woods, S., Walsh, B., Saksa, J., & McGlashan, T. (2010). The case for including attenuated psychotic symptoms syndrome in DSM-5 as a psychosis risk syndrome. *Schizophrenia Research, 123*(2–3), 199–207.

Chapter 3
Conceptual Engineering of Medical Concepts

Elisabetta Lalumera

Abstract There is a lot of conceptual engineering going on in medical research. I substantiate this claim with two examples, the medical debate about cancer classification and about obesity as a disease. I also argue that the proper targets of conceptual engineering in medical research are experts' conceptions. These are explicitly written down in documents and guidelines, and they bear on research and policies. In the second part of the chapter, I propose an externalist framework in which conceptions have both the explanatory power of psychological concepts and that of semantic concepts. It is likely, however, that human activities and practices distinct from medical research, and regulated by different practices and epistemic rules, call for different targets for conceptual engineering. I conclude by indicating an open agenda of problems for philosophers of medicine interested in conceptual engineering.

Keywords Cancer · Classification · Values · Medicine · Nosology · Concepts · Conceptions · Obesity · Experts

3.1 Introduction

There is a lot of conceptual engineering of medical concepts going on, both in medicine and the philosophy of medicine, but it is not usually recognized as such. This chapter aims to substantiate this claim through two case studies, cancer and obesity. I would also like to extract a methodological point about the conceptual engineering of medical concepts, namely that the concepts that can and should be engineered in medicine are, in the vast majority of cases, experts' conceptions. Experts' conceptions are explicitly written down in classifications and guidelines, they are epistemically accessible and public, and they have an impact on policies, research and cure. The considerations I will develop in favour of conceptions in this

E. Lalumera (✉)
University of Bologna, Department for Life Quality Studies, Rimini, RN, Italy
e-mail: elisabetta.lalumera@unibo.it

M. G. Isaac et al. (eds.), *New Perspectives on Conceptual Engineering - Volume 3*, Synthese Library 515, https://doi.org/10.1007/978-3-031-98535-5_3

case here may hopefully shed light on the issues of the feasibility of conceptual engineering, also known as the "implementation challenge", and of pluralism in conceptual engineering. The chapter is structured as follows. In Sects. 3.1–3.3, I illustrate how cancer and obesity, have been redefined in medicine, because of some specific goals. In Sect. 3.4, I argue that these are cases of conceptual engineering, and propose a semantic framework including conceptions. The above-mentioned methodological point on the conceptual engineering project from medical cases, before summing up my conclusions in Sect. 3.5.

3.2 Classifying Cancer

In this and the following section, I will illustrate two ways in which the concept of cancer is and has been, modified in medicine and, partially, in the philosophy of medicine, with a view to epistemic but also non-epistemic goals. In Sect. 3.5 I will explain that we also need the more fine-grained notion of conceptions of cancer to account for ameliorating changes—but for now, the expression "concept of cancer" will do. Let me start by reporting a few facts about what the concept is of.[1] The concept of cancer as it is used now in the medical community refers to more than 100 different diseases, characterized by uncontrolled growth of abnormal cells that can take place in many organs of the body. These cells typically pile up in an unstructured mass (tumour), and sometimes also spread to other parts of the body (malignant tumour). A very broad distinction is made among three kinds of cancers: depending on the type of tissue they begin in, these are carcinoma (a cancer that begins in the skin or cover tissues of organs), sarcoma (begins in bone, cartilage, fat, or muscle), and leukaemia (which starts in blood-forming tissue). Cancer may end up in organ failure, long illness, and premature death. Chemotherapy, radiation therapy, immunotherapy, and surgery are now the main interventions available. The choice among such options depends on the site (where the cancer originated), the grade (how the cells look), and the stage (how far it has progressed).

This explains why a good classification of cancers is important on the practical side. Different kinds of cancer respond differently to interventions, and to diagnostic procedures, so to know more about kinds of cancer is to be able to do more about them. How many kinds of cancer there are and how they are differentiated is also crucially relevant to epidemiology, to drug development, to diagnostic and therapeutic solutions based on artificial intelligence, and, last but not least, to institutional and private healthcare policies, and research funding. There is therefore constant and intense scientific work being carried out behind cancer classification, which is regularly updated in the light of new evidence. The official one, issued by the WHO—unofficially called "the Blue Books"—takes primary site as the

[1] For an updated reference on cancer and cancer treatment, see Russo et al. (2021). An important monograph on cancer from the point of view of the philosophy of science is Plutynski (2018).

organizing principle. Now there are 11 volumes in the series, including *Thoracic Tumours* and *Digestive System Tumours*, and so on (Bosman et al., 2010; Travis et al., 2015).

These classifications are used worldwide, but they are also frequently challenged by experts in different subfields of medicine. An open issue is whether, and if yes then to what extent, classifying cancers according to their molecular and genetic profiles could be feasible, and better than what we have now. Currently, genetic profiles merely integrate traditional classifications, but not uniformly across them—for example, we have robust molecular profiling notably for breast cancers, but not for digestive system cancers or for prostate cancers (about which I will say more below) (Tsang & Tse, 2020; Varghese, 2015). Diagnostic imaging data may also contribute to changing the morphology-based WHO system, especially when it comes to staging and grading, and are partially already doing so (Purysko et al., 2019).

In general, one could be tempted to think that the more we get to know, the more subtypes of cancers, or new cancer subconcepts, will enter the classifications; it is just a matter of time—and the more concept-splitting, the better. In fact, things are not so simple. We saw above on how many levels the decision of lumping together cancer types or, as more frequently happens, splitting a previous unified concept into two or more subtypes may matter. What should count as a new cancer subconcept? The problem is not only epistemic, but also practical, and even ethical. Suppose that a new proposed disease entity, let us say a glioma, a kind of brain tumour, can only be individuated through a very complex and expensive diagnostic procedure (for example, by combining histological description, molecular tests, and analysis with bioinformatics). The new concept would make brain cancer diagnosis more precise. However, most medical communities and practitioners worldwide do not have access to very complex and expensive diagnostic procedures. This technological disadvantage would not prevent them from understanding the new concept and from using it, for example, when discussing post-diagnosis treatment plans. However, neither would they apply the new concept in their everyday diagnostic activities, nor in their research. According to some, the introduction of the new concept would thus increase the already existing healthcare disparities. The general point here is that precision and worldwide shareability are two desiderata for cancer concepts, but they may conflict. Should we require a repertoire of cancer concepts, so to speak, that is "available to all through affordable tools and acquirable knowledge? Or should we embrace complex, detailed (and thus less affordable) genomic tools in this exercise?" (Salto-Tellez & Cree, 2019).

One solution could be to have a plurality of repertoires for cancer concepts, or a plurality of classifications—one for cutting-edge research, one for everyday clinical care and healthcare reimbursements, and so on. To put it simply, prioritizing research results in increasing cancer subconcepts; worrying about healthcare disparities resolves in sticking to common denominator nosologies; and caring more about the practical side of diagnosis and treatment involves conservativism, rather than frequent revision, of cancer concepts. But even within research, a classification that favours the goals of epidemiology is likely to admit a new disease entity individuated by its causes, whereas one that favours the goals of drug development would prefer splitting cancer concepts according to response to treatment (Richiardi et al., 2017). How to choose, and what to prioritize? The main worries with having a plurality of cancer repertoires, however, are that it would then the issue of who coordinates them and how (an epistemic cost), and it would definitely raise the epistemic risk of miscommunication among stakeholders (researchers, clinicians, patients, and institutions)—and communication between them is precisely one of the main aims of having a cancer repertoire at all. Possibly for these reasons, as we have seen, the WHO classification has split down cancer subconcepts in a conservative way. However it might be in the future, the choice here is not just evidence-based, but also philosophical in nature, as cancer concepts are carved out with epistemic goals, but also non-epistemic values in view (Plutynski, 2018).

This latter consideration points to a question in social epistemology, on the nature of the expertise required in concept splitting. In fact, there is some discussion going on about how diverse, specialty-wise, the community of medical experts should be when it comes to cancer classification. Should pathologists still have the last word, as they always did? And what about oncologists, molecular biologists, epidemiologists, and imaging specialists? Pathologists know the differences among different kinds of cancer cells, but other expertises may be relevant. The genetic profiling of cancer cells is becoming increasingly possible and useful, so there is a reason why molecular biologists should be involved. Epidemiologists know how frequent a specific tumor is in a population, and how it varies with time, so they may be involved in adding the information that that tumor is rare, or that it ceased to be rare. Oncologists know about the response of cancers to therapies, and they might want to claim, for example, that two genetically or histologically different kinds of cells are classified as the same tumour because they respond exactly to the same treatment. Imaging specialists, such as radiologists and nuclear physicians, are involved in cancer diagnosis and staging, so they know how early different kinds of cancers can be detected, and how they evolve (Salto-Tellez & Cree, 2019). Professional philosophers of science and medicine are, for the moment, not actively engaged in this process—though arguably the discussions going on in medical journals are philosophical.[2] It is plausible that including diverse medical specialties in the revision of cancer nosology, would result in different cancer subconcepts. Whether the contribution that each one can bring is purely epistemic (different knowledge about cancer), the decision about inclusion is based on social, political, and practical reasons internal to the medical research community. This is

[2] Readers familiar with the philosophy of psychiatry may have noticed the similarity between issues about classifications of mental disorders and cancer classifications. Differences are also striking, however. To put it simply, the community of specialists in mental conditions is less professionally and methodologically homogeneous than the community of cancer specialists, and psychiatry is much more contested than oncology (or medical imaging, or molecular biology) in its epistemic credentials. Partially as a consequence, philosophers are relatively more involved in discussions about psychiatric concepts, their definition, and subtypes. This, however, is not the place to develop this point further. See Rachel Cooper's chapter in this volume.

another aspect in which the concept of cancer is much less evidence-based than it may seem, prima facie.

3.3 Demarcating Cancer

Differentiating among subtypes of a superordinate concept, however, is just one of the two pillars of an adequate classification system. A superordinate concept is completely specified when all its subtypes are differentiated one from the other, but also—and maybe primarily—when we know how to fix the boundaries of the concept itself. To put it simply, telling apples from pears should come after, or at least go together with, describing what counts as fruit. In the case of cancer, it is crucially important to demarcate what cancer is from what it is not, and from what is not yet, namely from benign tumours and cancer precursor lesions, such as hyperplasia, dysplasia, and *in situ* tumours.

Before illustrating this demarcation problem, it is worth reminding ourselves in this case too that the importance of discriminating cancer from non-cancer and not-yet cancer is epistemic, but also existential and ethical. Whatever the prognosis is, a diagnosis of cancer is often a life-changing event for the person who receives it. A diagnosis of cancer may turn a person into a chronic patient. It may bring fear, stress, anxiety, and deep changes into one's life projects, expectations, or even values. This tends to happen irrespective of how much the person knows about medicine, or even specifically about cancer—along with the scientific literature, such an effect is well described in a quite well-known book, which collects the experiences of a group of cancer specialists and bioethicists who became cancer patients (Dresser, 2012). Thus, on the one hand, it is good to identify cell mutations—and diagnose the disease—soon, because cancer treatment is more effective when it starts in the earliest stages. On the other hand, however, it is bad to identify cell mutations *as cancer* too soon, as overdiagnosis and overmedicalization of non-risky conditions are harmful to the patient for the reason we have given. Subjective harms resulting from overdiagnosis and overmedicalization can turn into clinical risks (Hofmann, 2014). Moreover, when considering healthcare fairness, or the just distribution of resources, they are also ethically wrong, because they divert resources from where they may be needed to where they are futile (Bosman et al., 2010; Lysdahl & Hofmann, 2020).

Steering a balance between these two desiderata (soon is good, but not too soon is also good) is one of the main philosophical problems posed by the concept of cancer.[3] Cancer poses many problems, such as how to cure it, how to explain its

[3] From the point of view of philosophy of science, cancer raises questions including: Is cancer a single disease, despite its heterogeneity in causal pathways and genetic profiles? We have competing general theories of cancer in medicine, is this situation epistemically good? Does mechanistic explanation apply to the case of cancer? Do holistic approaches to explanation fare better? How to define what counts as carcinogenic? For an overview, see (Plutynski, 2018).

formation, how to make diagnosis and treatment affordable to everyone, and so on, but this one can be singled out as philosophical in that it involves a decision on the boundaries of the concept, which is underdetermined by evidence, and it should be based on an assessment of epistemic and non-epistemic values—just as we noticed for the case of splitting cancer, in the last section.

Why are the boundaries of the concept of cancer underdetermined by evidence? First, because arguably the concept of cancer is semantically vague, just like the concept of a heap—one grain of rice does not make a heap, one million grains certainly does, but in a range of intermediate cases, there is no (knowable) fact of the matter as to whether the considered amount is a heap or not. This situation is frequent in medicine whenever the pathological condition can be represented by some of the values of a continuous variable. Think of hypertension, for example. Where does hypertension start? And is it a disease? Philosophers of medicine have noticed that in such cases, a line-drawing problem arises (Schwartz, 2007). Usually, the line is drawn by considering the relative risk of death or illness associated with various values of the variable in question. But also, the possibility and efficacy of the intervention (if it works), and even its cost-effectiveness (if it works well enough compared to its cost, evaluated on many levels), matter to the decision on where to draw the line. As the philosopher of medicine Elselijn Kingma (2015) made clear, those are prudential reasons, not facts; if this is so, then the line-drawing problem arising for many medical conditions may be counted as a case of "value intrusion" in a supposed realm of pure facts. Whether values intrude (a term that carries with it a derogatory sense), or whether they are intrinsic to medical research and practice, is a well-known open and big question in itself (Amoretti & Lalumera, 2021).

An objection can be raised at this point, going back to cancer. It points out that the concept of cancer is not vague like the concept of a heap, or the concept of hypertension, because one abnormal cell should count as cancer precisely because of the mechanism of its abnormal growth. The quality, not just the quantity, of cells matters when it comes to cancer, whereas this is not the case for (hyper)tension, or a heap. One response is that we would have no use for such a purely biological concept of cancer. Even if it were an epistemically good concept—so to speak—because it categorizes similar phenomena together, on the grounds of similarity of mechanisms, it would not be a clinically useful concept. A person with two abnormal cells in their lung and a person with metastatic lung cancer do not have much in common from the clinical, or indeed the existential, point of view. The concept of cancer we need in medical research and practice ought to have both a mechanistic or biological ground and clinical utility. This, however, brings us back to the core of the issue: that the concept of cancer ought to have both aspects and which one to prioritize in different contexts is a philosophical decision. The prostate cancer example that I will discuss just below will give evidence to this claim.

A slightly different way to frame the problem of drawing the line between not-yet cancer and cancer is to focus on the distinction between benign and malignant tumours, or proper *cancers*. The difficulty here is not just that of individuating a discrimination point in a continuum, but additionally that of choosing between different criteria for malignancy, namely between histological and clinical criteria. A little medical information is required here. Histological criteria, the traditional ones, are roughly what the pathologist can see. According to these criteria, benign tumours resemble the morphology of the original tissue and never infiltrate surrounding tissue; they might grow expansively but never replace other organs. Moreover, in benign cases there are no tumour cells in blood or lymphatic vessels; in epithelial tissues, any cellular proliferation is limited to the epithelial basal membrane; in mesenchymal tissues cells are typical and the mitotic activity is low or absent; and in lymphoid tissues, the histological architecture is preserved. Malignant tumours from the histological point of view are just the opposite: they infiltrate surrounding tissues or disseminate to distant organs (metastases); they disseminate tumour cells in blood or lymphatic vessels; in epithelial tissues, they invade the basal membrane; in mesenchymal tissues, they display atypical cells; and in lymphoid tissues, there is a change of the normal tissue architecture. Clinical criteria are about what matters to the patient's risk of disability or death. Clinically speaking, benign tumours never infiltrate surrounding tissues, never metastasize, never recur; they might have expansive growth but never replace other organs; and they do not usually affect general body performance. Complementary as they may seem, histological criteria and clinical criteria do not always overlap. It has been shown that they do not, at least in the case of low-grade prostate cancer, encapsulated thyroid tumours originating from follicular cells, low-grade papillary carcinoma of the bladder, intra-mucosal adenocarcinoma of the colon, *in situ* adenocarcinoma of the lung, and clear-cell tubule-papillary carcinoma of the kidney (Esserman et al., 2014).

Let us look in more detail at just the first example, low-grade prostate cancer. With advanced diagnostic tests such as PSA (Prostate-Specific Antigen) screening it is now easy to detect microscopic foci of prostate cancer in men. When PSA screening was introduced around 1990, there was a logarithmic increase in diagnosed cases, and the early treatment of young patients began; currently, most prostate cancers are diagnosed at the first stage—specifically, at the stage classified as Gleason 6 according to the traditional system. This is not entirely good, however. It was found that overdiagnosis of small prostate cancers had no significant positive effect on mortality. This is because it brought with it a diagnosis of tumours that, in virtue of their characteristics, would never be clinically manifest, that is, are unlikely to cause any harm to the patient—the so-called "indolent tumours".[4] Intervening on these tumours (with surgery, radiation, or other interventions) counts as an overtreatment or futile treatment. It means that it does more harm than good to the patient. It may cause anxiety, depression, and general loss of well-being, on top of the side effects of the interventions. Overtreatment is now a big problem in medicine, and it also raises considerations of equity and fairness—if a healthcare system provides futile interventions, it diverts resources from useful ones (Esserman et al., 2014) In light of these facts, around 2012–2016, the issue was taken up

[4] Not all small tumors are indolent, and in other cases an early diagnosis and intervention can save a life. What became clear in the last two decades, however, is that not all small tumors should be treated as soon as they are discovered.

and discussed by the urology and oncology medical communities in a series of consensus conferences of experts on prostate cancer. One point raised was semantic: should the label "cancer" be dropped altogether for these abnormal cell formations, which are most unlikely to cause harm? It was noted that fear of death resulting from cancer likely plays some role in overtreatment, and removing the scary label could reduce it (Carter et al., 2012). As a result, low-grade (specifically, Gleason 6) formerly prostate cancers were declassified, and it was proposed that they are not called "prostate cancer" anymore, but rather "indolent lesions of epithelial origin" (IDLE) (Epstein et al., 2016).

It is important to note here that the declassified cell formations count as malignant according to the histological criteria—they can extend locally beyond the prostate and invade nerves—though clinically they are not dangerous. In this case, the concept of (prostate) cancer was reshaped according to clinical criteria, which were preferred over histological criteria. Was it the only option available in the light of evidence? Surely not. Another option, favoured by specialists in the US, was (and still is) to change just the grading label—if it is low-grade cancer, let us call it Grade 1 cancer, and not Grade 6 (Epstein, 2022). But when the decision was taken, it was the best option in the light of the evidence, plus the accorded preference for clinical criteria, and considering the ethical principle of non-maleficence (or doing no harm) as the main goal. In these respects, the concept was engineered, or ameliorated, with a view to non-epistemic goals.

The philosopher of medicine Peter Schwarz argues for a similar conceptual change for the case of ductal carcinoma *in situ*, a small breast cancer, which can now be easily diagnosed during routine screening tests. The similarity consists in the conclusion that such lesion cancers should be declassified, just like small prostate cancers, because they are unlikely to develop into metastases, and therefore cause serious harm, whereas the harm of overtreatment is well documented (Schwartz, 2014). The difference with the case of the prostate I have just illustrated is that Schwartz's argument focuses on the concept of disease, and not on the concept of cancer. Adopting the theory of disease as dysfunction, defended by Christopher Boorse (Boorse, 1977, 1997), and balancing such conceptual insight with evidence on the prognosis and risk of ductal carcinoma *in situ*, he claims that this kind of malignant cell formation is not a disease, but a condition of increased risk. This does not mean, however, that the condition is clinically irrelevant. In fact, Schwartz holds that:

> the judgment of whether a condition is a disease is separate from whether it carries risks and whether there are treatments that can reduce that risk without imposing unacceptable burdens, cost, or new risks. In fact, simply describing [ductal carcinoma *in situ*] as a disease, if done without emphasizing the assessment of risk and benefit in choosing treatment, might interfere with helping individuals determine whether to undertake certain steps for risk reduction. (Schwartz, 2014, p. 997)

If being a disease does not equate with being clinically relevant, why bother about demarcating disease from non-disease, in this case as in others, then? Because Schwartz thinks—and I agree, without pursuing the point further here—that there are good arguments for viewing medicine as the science that deals with diseases. Some such arguments, in passing, involve considerations about what a just healthcare system should be, and what it should provide freely to citizens in a democratic society, that is, cure and care for diseases and not, for example, medical enhancement. If we want to ensure that limited healthcare resources are equitably distributed, we need a distinction between diseases and other undesirable conditions first.

To quote Schwartz's conclusion again:

> seeing medicine in this way relies on being able to draw lines between normal, healthy conditions and abnormal or pathological ones, and while there are complexities and challenges to such line drawing, it is still central, and essential, to medicine. (Ibid.)

3.4 Obesity as a Disease

Schwartz's conclusion above points out one reason why it is important—rather than merely intellectually engaging—to establish whether a condition is a disease or not. Diseases, unlike other undesirable conditions, such as poverty or aesthetically imperfect bodily features, are what medicine should aim at explaining, caring for, and curing when possible.[5] Another closely connected reason is that diseases generally (though not invariably) bring with them a social status, the so-called *sickness* aspect—conceptually distinguishable from the objective, bodily condition (disease proper), and from *illness* (the subjective state of feeling bad). The sickness aspect is a set of rights, permissions, or obligations, both at the personal and at societal level (Amoretti & Lalumera, 2020; Hofmann, 2002). A person with a disease can be granted sick leave from work, receive a pension or other benefits, and can be declared as legally non-responsible, or less culpable, if she commits a crime; she is eligible to receive reimbursements, or free treatments from doctors and hospitals, depending on the organization of the healthcare system of the country she lives in. At the societal level, when a condition is recognized as a disease, research funding can be allocated for studying it, healthcare facilities can be created to cure it, and policies can be implemented to prevent it. This is why fibromyalgia patients in the past, and more recently long-Covid patients, have been advocating the recognition of a disease status—a nosological entry, a proper diagnostic concept—for their condition (Callard & Perego, 2021); (Häuser & Fitzcharles, 2018); more examples, such as disposophobia and Asperger syndrome, can be found in the recent history of psychiatry (Cooper, 2014). At least in some cases, when a condition becomes a disease—when the concept of that condition becomes a disease concept—there are significant social advantages for affected individuals and groups.

[5] There is actually a debate going on about how to characterize the goals of medicine, of which this is just one of the possible positions. See, for example, Schramme (2017), Broadbent (2019), and Chap. 2.

There may also be undesirable consequences of the sickness aspect, though. It is sometimes the case that people with a socially recognized disease status are stigmatized, that is, perceived as different by non-diseased people, and excluded and marginalized based on this. Stigmatization is arguably both morally bad and clinically disadvantageous, as stigmatized people tend to internalize the negative beliefs that society projects onto them, with the result that they give up treatments and renounce healthcare facilities (Goffman, 2009). However, not all diseases are stigmatized—think of the difference between the cancer patient, who tends to be represented as a fighting hero, and the depressed patient, who is definitely not (Corrigan & Bink, 2016; Seale, 2001). Therefore, it is debatable whether stigmatization is merely a consequence of the disease label, or if other variables are in place. We will see this question exemplified in our second main example in this chapter, concerning obesity.

As is well known, obesity has been steadily increasing worldwide in the last 50 years, and now it represents a serious threat to life expectancy in many countries. In 2008, the World Obesity Federation, a medical scientific society based in the United States, published their first position paper about the status of obesity as a disease, including the following explicit definition (italics mine):

[O]besity is a multi-causal chronic disease recognized across the life span resulting from a long-term positive energy balance with the development of excess adiposity that over time leads to structural abnormalities, physiological derangements, and functional impairments. The disease of obesity increases the risk of developing other chronic diseases and is associated with premature mortality. As with other chronic diseases, obesity is distinguished by multiple phenotypes, clinical presentations, and treatment responses. (Allison et al., 2008, p. 1161).

The position has been recently reaffirmed, and the paper in the last version is supplemented with evidence that other scientific societies have endorsed analogous definitions of obesity as a disease in the last decade—specifically 23 of them, located worldwide, and including institutions as diverse as the World Health Organization, the Canadian Medical Association, the European Association for the Study of Obesity, and the Korean Society for the Study of Obesity (Jastreboff et al., 2019, Table 1).

Let me briefly go through the line of argument adopted by the panel of experts of the World Obesity Federation to arrive at their conclusion. First, they assess the possibility of adopting what they called "the scientific approach", namely establishing whether obesity is a disease based on objective facts. However, they argue that while there are objective facts of the matter about what obesity is, there are none about what a disease is. In particular, there is a lack of a clear, accepted, and scientifically applicable definition of "disease" that allows one to objectively and empirically determine whether specific conditions are diseases.[6] Then they contemplate the possibility of considering the statements of the majority of relevant

[6] The authors of the position paper consider various definitions of disease defended in the contemporary medical literature or endorsed by scientific associations, such as the American

experts and societies in the field, which they call "the forensic approach". The problem with this approach, they conclude, is that it is constitutively insufficient to tell us what is good and right, as different from what is consistent with mainstream opinions. Finally, they decide in favour of what they deem a "utilitaristic approach" to the question of whether obesity is a disease, according to which:

> conditions that produce adverse health outcomes come to be considered diseases as the result of a social process when it is assessed to be beneficial to the greater good that they be so judged. Such judgments about likely benefit to the greater good are utilitarian judgments that may take empirical input but must also assume certain values. (Ibid.)

Leaving aside the philosophical naivete of the terminology chosen by the World Obesity Federation, the three approaches they consider map onto recognizable philosophical strategies for the task of deciding whether obesity is a disease—and therefore of engineering the concept of obesity, by adding, or not adding, a disease component. The first strategy is what we have seen in the previous section when presenting Schwarz's proposal about ductal carcinoma *in situ*. Schwartz started from a definition of what a disease is (Boorse's definition)—a conceptual fact—and then argued that the small breast tumours in question are not diseases, on the ground of such a conceptual fact. This first strategy consists in answering the question of whether the concept of obesity should be engineered into a disease concept, by pointing at the superordinate concept of disease as evidence. The "forensic approach", on the other hand, is analogous to approaching conceptual engineering as tracking socio-linguistic changes in the use of a term. In other words, this second strategy consists in answering the problem of whether the concept of obesity should be engineered as a disease concept by using sociolinguistic data as evidence (in particular, data about how the concept is used in documents of various institutions and scientific societies). Finally, the "utilitaristic approach" is a ~~bad~~ name for the idea that concepts can be ameliorated with a view to non-epistemic goals. It consists in grounding the decision about the concept of disease on considerations of how it would affect the well-being of people with this condition, a fair distribution of research funding, etc. We will return to these issues in the next section.

What are the good effects that would follow from changing the concept of obesity into a disease concept? Here is a list, extracted from the World Obesity Federation paper and statements by the other societies mentioned in the 2019 document. The change would solicit more resources on the treatment and research of obesity; encourage more health professionals to view treating obese patients as a vocation worthy of effort and respect; shift public perceptions of obesity away from the view that obesity is a lifestyle choice or aberrant behaviour toward a chronic disease model with individual and population-level determinants; reduce the stigma and discrimination on many obese persons; support policies of regulation for the food industry; facilitate efforts to regulate the marketing of unhealthy foods and drinks to children and adults; and promote structural and environmental changes

---

Medical Association. As these mostly recapitulate the debate in the philosophy of medicine about the concept of disease, I refer to (Stegana, 2018; Amoretti & Lalumera, 2021).

in workplaces, schools, and communities to prevent obesity and promote physical activity and healthy eating. In sum, most of these good effects have to do with the social aspect of the concept of disease, namely with the rights and obligations that it carries with it (Jastreboff et al., 2019, p. 9).

There is, however, no uniform consensus about whether changing the concept of obesity into a disease concept really amounts to ameliorating it. Or at least, if there is a good consensus within medicine, things are less clear in other scientific and non-scientific communities. Let us look at the disagreements within medicine first. Some scientists claim that obesity cannot be a disease, as its operationalization is flawed. The condition is mainly assessed via the body mass index (BMI), the value obtained from a person's weight in kilograms divided by the square of the height in metres. It is widely recognized that people with the same BMI may be very different, depending on muscle mass, bone density, fat distribution, and other variables; therefore, it is now often claimed that BMI does not individuate phenotypes that are precise enough for research and clinical needs. It is true that scientific societies, such as the World Health Organization, avoid defining obesity via BMI, but in fact, BMI remains the way to measure it in most contexts. For this reason, critics of the proposal that obesity is classified as a disease think that obesity should be assessed in other ways, namely by checking the overall health status of the person. But if this is so, then obesity is not a disease condition per se, but rather it is a disease condition just for those people who are not in good health—which is either trivial or circular (Müller & Geisler, 2017). A possible reply may be that at least some values of BMI are correlated with bad health outcomes beyond a reasonable doubt; however, the methodological point remains open.

A parallel concern about obesity and BMI is voiced by some sociologists and philosophers. It is claimed that "the conflation of measurement with a formal disease label" would encourage stigma rather than reducing it in a society where the idea that bodily appearance is a matter of self-discipline and virtue is deeply rooted (de Vries, 2007) (Gutin, 2021). In fact, the modification of the obesity concept as a disease concept has not yet been tested for its social effects; it is an empirical question whether its implementation will be beneficial or not—as I said above, the association of stigma with the disease label is variable. Moreover, as obesity is correlated with socio-economic status and ethnicity, there is the worry that a disease label would reinforce racism and discrimination. At least in the United States, obesity is prevalent among African Americans and Latinos in general, and in black and Latina/o women and children in particular. The worry is that the obesity diagnosis would strengthen the association of black and brown femininity with pathology, decadence, lack of self-control, and irrationality (Sanders, 2019). To this worry, it has been replied that the modified concept would rather help channel resources for health prevention to disadvantaged socio-economic communities—but again, the issue is empirically open.

A further criticism from outside medicine refocuses the problem of the concept of disease involved in the obesity as disease proposal. The philosopher of medicine Bjorn Hofmann (Hofmann, 2016) points out that the conceptual change in question has been decided on pragmatic grounds; that is, it is purely value-based—this is what the position paper of the Obesity Society admits, as we saw above. Starting from this premise, and by testing the obesity-as-disease claim with various philosophical concepts of disease, Hofmann argues that the claim is justified only if a social concept of disease is endorsed. This is the idea that social norms of desirability, advantageousness, and conformity to present-time ideals of efficiency, success, or even morality are conditions for being a disease. As examples of diseases in the social sense, Hofmann proposes homosexuality as it was classified until about 1980,[7] and attention-deficit/hyperactivity disorder (ADHD) now. The upshot of his argument is that there are ethical risks in promoting and applying a social concept of disease, concerning again stigmatization, ableism, and discrimination. If Hofmann is right, then changing the concept of obesity into a disease concept commits the medical community to an undesirable concept of disease:

making obesity a socially defined disease may well turn out to be a wise health policy, improving population health. Especially if we address the social determinants of obesity, such as the food supply and marketing system, disparities in socio-economic status, transportation system, and consumption growth. But to rely too heavily on healthcare may also turn out to be an ineffective and harmful way to solve socially defined challenges. (Hofmann, 2016, p. 97).

Hofmann's point in the quote above can be put as thus: when a condition is classified as a disease (as in the case of obesity) or de-classified as a disease (as in the case of homosexuality) on value grounds, i.e. to fix social problems connected with the condition, these social problems become healthcare problems, and the risk is that their political and structural dimensions are somehow neglected. To put it very simply, if it's a disease, then doctors will take care of it—and political institutions can well ignore the issue. This is a point worth considering whenever the concept of disease is at stake, and it adds up to the complexity of the engineering of medical concepts.

3.5 On Conceptual Engineering

Are you convinced that the obesity as disease concept is better than the obesity as risk condition concept, or is it the other way around? And do you think that histological criteria for cancer should continue to guide classification and demarcation? Whatever the answers may be, my aim here was not to give you the necessary details to arrive at them. My case studies, cancer and obesity, were meant as evidence for the claim that there is conceptual engineering going on in medicine, and in the philosophy of medicine. Conceptual engineering can be defined as "the activity of

[7] More precisely, the American Psychiatric Association voted in 1973 that "homosexuality ... by itself does not constitute a mental disorder" (Bayer, 1987) p. 128). But ego-dystonic homosexuality remained as a disorder in the revised version of DSM-II, published in 1974. The diagnostic entry was definitively removed in DSM-III, issued in 1980. That is why it can be said that homosexuality was "classified" as a disorder until about 1980.

assessing and developing improvements of our representational devices" (Cappelen & Plunkett, 2020, p. 133). In the examples, we saw how the medical community, or relevant subcommunities within it, proposed to make ameliorative changes to the concepts of cancer and obesity. They faced the question of what concepts of cancer and obesity, respectively, they should have—a normative question (Dutilh-Novaes, 2018; Simion, 2017). In particular, Section 3.1 about cancer classification ended up by showing the different and alternative values that may shape the decision to create a new subordinate concept of cancer—a new nosological entity—rather than a fully implemented change. In Cappelen and Plunkett's (2020) terms again, this is a case of non-purely epistemic amelioration, as we saw that new cancer subconcepts are not just subject to the constraint of being more precise, but also more useful, less harmful to patients, and more advantageous for the healthcare sector as a whole. In the section about cancer discrimination, again, we saw a case of conceptual amelioration in view of a practical and ethical goal, namely, to avoid overtreatment as well as clinical and psychological harm to the patient. This was successfully implemented. Finally, the discussion of obesity showed a partially successful conceptual engineering project, again driven by practical and ethical goals.

The discussion on the cancer case and the obesity case could be conducted on different levels. First, they are evidence for a claim that many philosophers of science and medicine have argued for in recent decades, namely that science in general, and medicine in particular, are value-laden in an ineliminable way (Amoretti & Lalumera, 2021; Douglas, 2009; Longino, 1996; Rudner, 1953). Second, they also testify to the role of consensus conferences and expert meetings in the making of medical knowledge—medicine is not just evidence-based, but it is relevantly based on people's agreement about practical and ethical choices (Fanti et al., 2019; Solomon, 2016), (Stegenga, 2016).

Moreover, they might be taken as evidence against a tendency within the contemporary philosophy of medicine, namely that of discarding as irrelevant any discourse about concepts in medicine—viewed as the old tradition of conceptual analysis, the search for necessary and sufficient conditions for the application of a concept, that is, a definition immune to counterexamples (Fuller, 2018; Lemoine, 2013; Sholl, 2020). I believe that the cases I presented show that the discussion about concepts is alive and kicking in medicine, and it brings about changes. So, it is something philosophers of medicine should pay attention to, and possibly take part in, with discussion and arguments. Authors of the anti-conceptual tendency, however, are right in pointing out that purely descriptive conceptual analysis—with no amelioration in view—is not part of medical research, and this may count as a reason for downgrading it in the philosophers' agenda as well.

In this chapter, however, I intend to illustrate how the cases of conceptual change in medicine I presented bear on the ongoing philosophical debate on conceptual engineering. The rest of this section will then be dedicated to this task. I will start with the questions: What are the representational devices that we can ameliorate and change in the cases I presented? Are these the ideal or proper objects of conceptual engineering in general? This will bring me to illustrate the metasemantic framework for conceptual engineering I favour. To anticipate, I distinguish between psychological concepts, semantic concepts and conceptions, and the latter are supposed to do most of the explanatory work—I believe that conceptual engineering in medical cases like cancer and obesity is best understood as modification of experts' conceptions. I also tentatively propose that other domains of discourse call for other frameworks, some of which do not include conceptions in a central explanatory role. The engineering of the concepts of concept and conception, however, is not my main goal here, and this section has the more modest one of illustrating one possibility, to be explored further elsewhere. So, let us get down to details.

On the one hand there are psychological concepts. The core idea here is that concepts are constructs that explain people's behaviour in a wide set of cognitive tasks including recognition, categorization, inference, imagination, etc. Psychological concepts are whatever people use by default to identify and categorize individuals or kinds of things—not just beliefs, but also prototypes, exemplars, theories, etc. According to this characterization, a psychological concept of C is a functional kind that can be realized by different psychological entities (Isaac, 2020; Lalumera, 2010; Machery, 2017). My psychological concept of obesity, for example, may include exemplars of obese persons I have stored in memory, some beliefs about body-mass index thresholds for obesity, as well as the belief that obesity is a contested. Psychological concepts are highly variable among different people (my concept of a kind of thing can be quite different than your concept of the same kind); they can also be diachronically variable within-subject (my concept can become richer or change because of my experiences, testimonial knowledge, and all possible changes in my level of familiarity with the kind in question and expertise).

On the other hand, there are semantic concepts. These are constructs that explain the truth of categorial judgments and the validity of inferences. The semantic concept of category C is what truly applies to all and only things that are C. Its intension, definition, or constitutive principles may diverge, and often do, from what people use fallibly to identify or recognize C. A semantic concept is also (according to many), the meaning of the word (or words in various languages) that the concept expresses. Traditionally, there are three main alternative accounts of how a concept applies to (or stands for, or represents, or is true of) something. Descriptivists hold that the definition of C (the intension), determines the extension of the concept of C, including of all and only things that fit the definition (Evans & Altham, 1973). One can relax the point on definitions a bit, and propose that a set of constitutive principles implicitly individuates C (Peacocke, 1992). Direct reference theorists deny all this, and insist that what makes something the concept of C is a causal (or more generally, reference-fixing) link with C.

In a somewhat metaphorical, and yet important sense, semantic concepts are not in people's minds, but rather they are grasped by people's minds, sometimes fully and some others partially (and I will say more on this later). For example, I fully grasp the concept of a prime number (I know the definition that generates all and only prime numbers), whereas now I do not fully grasp the concept of, say, the limit of an arithmetical function—though I used to, when I was in high school

(Peacocke, 1992, 1998). Traditional conceptual analysis can be characterized as the quest for semantic concepts so conceived—usually taking the form of a quest for correct definitions (Laurence & Margolis, 2003).

Should conceptual engineering be directed at changing psychological concepts, or semantic concepts? Let me mention here that one could skip the question altogether and remain neutral, proposing formulations like the one I mentioned above, that conceptual engineering is aimed at ameliorating "representational devices" (Cappelen, 2018; see also (Burgess & Plunkett, 2020; Isaac, 2021) for a critique). In the specific context of medicine and perhaps more generally science, however, I believe that the choice of what to target makes a difference, as I will explain below. So let me dismiss neutrality, and briefly assess the pros and cons of the above-mentioned alternatives.

The main reason in favour of psychological concepts as proper targets is that conceptual engineering aims at real-world changes of people's behaviour, and psychological concepts are what people actually use in the real world when they judge, discriminate, make plans, and so on (Isaac, 2021). One reason against engineering psychological concepts concerns the feasibility of the project—can it really be accomplished? This kind of question in its general version is known as the implementation challenge for conceptual engineering, which concerns semantic concepts, for different reasons.[8] Modifying psychological concepts might turn out to be difficult if feasible at all, considering that they are highly variable intersubjectively and intra-subjectively over time, and they are difficult to get to know about. We would need an empirical study to find out what is the concept of obesity that a group of people employs, and after that, the mechanisms of concept formation and change to intervene on are many and complicated. as feasibility correlates with scope—given a small group of people, in a limited time, it is possible to investigate empirically what their psychological concepts are and to study how to intervene (Jorem, 2020).

The second reason against concentrating on psychological concepts is not practical but methodological, and it can be turned into the main reason in favour of semantic concepts as targets of conceptual engineering. With conceptual engineering, we aim at altering the truth value of categorical judgments, such as "this is cancer" or "this is a woman", or "this is a disease" applied to particular cases, sometimes new cases (Flocke, 2021). To do that, we need to change what truly counts as cancer or women or disease, that is, the semantic concept (intension and/or extension), or the meaning of the associated word. The problem with this view has been put in terms of irrelevance: even if we could change the meaning of our target concept, people might go on using their old psychological concepts all along. Precisely because semantic concepts are not in our heads—as externalism requires[9]

[8] The implementation challenge is discussed in a number of recent works, including (Cappelen, 2018; Deutsch, 2020; Deutsch, 2021; Koch, 2021a; Pinder, 2021; Pollock, 2021; Riggs, 2019).

[9] Classical references for meaning externalism of various kinds are (Burge, 1979; Kripke, 1972; Putnam, 1975).

-, changing them can be irrelevant to how we reason and act (Koch, 2021b). A different but related worry has to do with the metaphysics of semantic concepts; as David Chalmers observes, it's not obvious that you can engineer a new resident of the third realm (Chalmers, 2020).

Still, the most ambitious claim that conceptual engineering should be addressed at changing both how people categorize and reason, and what counts as a true categorization judgment containing a certain concept is attractive, even in the simplified version I have just sketched—if we want to change our concepts, we need both the semantics and the psychology.[10] What is to do, then? In previous papers (Lalumera, 2014) I argued that the notion of conception can do some useful theoretical work here (and elsewhere). Conceptions may be the missing link. Though there are diverse notions of conception in the philosophical market, this is the one I favour: conceptions are what people have in mind when they possess a (semantic) concept. They are internalizations of semantic concepts. This characterization of conceptions comes from Christopher Peacocke's work on Fregean concepts (Peacocke, 1992, 1998), and its most recent defence is articulated by Sarah Sawyer (Sawyer 2018; Sawyer, 2020) (with a difference I will spell out right below here). Conceptions are part of externalist semantic frameworks, and they are often characterized as sets of beliefs about a category. I prefer the expression "what people have in mind" because it leaves open that nonconceptual and yet semantic content (such as perceptions) be included (this was an essential part of Peacocke's project). Conceptions stand to concepts as many to one. They have normative properties (they can be applied correctly or incorrectly), but they do not fix the extension of a concept. They can be fully explicit, and take the form of a theory, a list of principles, or a definition, but they can also be very sketchy—for example, my conception of inflation may just amount to 'a rise in prices that economists study and that is bad'—and implicit, possibly for example, one's conception of everyday objects like chairs and cups, or game, if Wittgenstein was right (Wittgenstein, 2010).[11]

The advantage of buying into conceptions is that one can say that different people, or the same person in different times, can have different conceptions, and yet possess and use the same concept. You can have both variation and conceptual sameness, disagreement and topic continuity. You can also make sense of the intuition that everyone's conception of something is wrong, and yet there is something all such conceptions approximate. If you're a causal theorist, in order to

[10] See (Koch, 2021b) for the same conclusion with detailed arguments. See also (Fischer, 2020).

[11] The motivations for including conceptions in one's metasemantic framework are similar to those for adopting a dual-aspect theory of concepts, for which I point again to (Koch, 2021b). The difference between conceptions and one of the two aspects of concepts envisaged by the dual theorists are subtle. A quick way to illustrate it could be that conceptions in my view have normative properties, whereas the cognitive psychological parts of dual aspect concepts do not. An extensive discussion of this issue, however, would lead us too far from the main focus of this chapter.

account for sameness and continuity you would need to defend the view that having the same concept, for people with different conceptions, comes with a causal link with the relevant category (Sawyer, 2020). If you are either a descriptivist or a dual theorist, on the other hand, you would opt for some variant of the idea that different people can understand the constitutive principles of a concept differently -, for example partially, implicitly, or deviantly.[12] Yet, insofar as they fulfil the minimal conditions for concept possession, they all refer to the same category, individuated by the definition or by the constitutive principles. This latter is the alternative I favour, but I will not defend it further here, as it does not make any relevant difference to my main claim in this chapter (which is that at least in some relevant cases, conceptual engineering in medicine is engineering of experts' conceptions). Finally, conceptions are in people's minds, so they are not causally inert as semantic concepts are usually considered.

Since psychological concepts are in people's minds, too, I still need to specify the difference between psychological concepts and conceptions in my view, before closing this metasemantic interlude. I take it to be mainly a difference in explanatory function. Psychological concepts explain default categorization, while conceptions, like semantic concepts, explain truth and correctness of judgments.[13] These can and do come apart—it may well be that what I use for the default categorization of, say, gold is something of the form "shiny and bright yellow metal", and yet this is not what I take to be the definition or the constitutive principles of gold (however incomplete these might be in my mind). From the epistemic point of view, then, a person's conception of C can be inferred from the judgments containing C she takes as true. Note that there is room for a distinction between truth and correctness here: someone whose conception of fish includes whales will correctly apply her conception of fish to whales, but she will produce a false judgment. So, unlike psychological concepts, conceptions can be applied correctly or incorrectly—they have normative properties. Unlike semantic concepts, however, conceptions are part of people's mental furniture and thereby they matter to what single agents do and think—so that the irrelevance objection to the claim that they should be the targets or conceptual engineering loses at least part of its grip.

I finally arrived at proposing that what goes on in medical research in cases like those I discussed—cancer and obesity—is conceptual engineering targeted at expert conceptions, or more precisely, at the conceptions that experts agree on. Here are the explanatory virtues of experts' conceptions in this context. They are always explicit; they take the form of outright definitions or constitutive principles. These are proposed and discussed in consensus conferences and published in articles and guidelines. Therefore, they are not ineffable, but rather epistemically easy to reach; and they are public. These two characteristics make medical experts' conceptions something it is possible to act on. Also, they are implemented in actions—in

[12] And yet they are entitled to them -this view is articulated in (Scharp, 2013).

[13] This might apply to the cognitive contents of dual aspect theories of concepts as well, see footnote 11 above.

research, care, cure, and healthcare policies. So, they are something it is worth acting on (again, no irrelevance objection lurks in here). It can still be objected that on a strictly externalist view (not a socio-externalist one) conceptions per se do not determine meaning, there is always a logical gap between use and meaning. So when experts' modify their conception, it is not the case that the extension of cancer or obesity ipso facto changes. Moreover, after all, experts can be wrong. I have a concessive reply to this worry. In the particular case of science and medicine, the correctness of experts' judgments ("Covid-19 is caused by a virus", "Pluto is not a planet") is generally considered to be a proxy for truth. Experts' conceptions do not fix what's true, but still—on this strongly externalist view—they fix a socially recognized semantic standard of correctness. A change in publicly recognized standards of correctness can be good enough for the practical purposes we aim at—this is how we can obtain semantic change, albeit in this proxy form, in an externalist framework.[14] Notice again that this proximity of correctness to truth is a peculiar feature we ascribe to scientific judgments, and it need not generalize to other less hierarchically structured knowledge communities.

Again, a few words are in place to recap why neither psychological concepts nor semantic concepts would be the ideal target of conceptual engineering here. It is surely true that how people think about their conditions (psychological concepts) has an impact on what counts as a disease, on what counts as urgent research, and on healthcare policies—to cite a few aspects only. Consider the case of homosexual activism that led to the (slow but definitive) de-pathologization of homosexuality in the late Seventies of last century. People's ways of thinking did make a change in that situation. However—I submit—they made a change by altering experts' conceptions, that is, what figures in nosologies, guidelines, research projects, and policies. It was only when a new conception of disorder and homosexuality was adopted by the experts, that real-world effects became regular and stable on the life of homosexual people. The hierarchical distribution of knowledge and power in medicine explains this fact. This by no means denies the possibility that experts' decisions, and therefore experts' conceptions, be arrived at in participatory and inclusive ways, so that what people think is given a recognized epistemic and political role in the engineering process. In fact, the tendency now is to advocate for more inclusive expert panels in medicine, and philosophers of science are very active in the project (Bueter, 2019; Tekin, 2022). My point here is just that changing people's psychological concepts alone—granted that this is possible, for the sake of the argument—would not be sufficient in the context of medical research, for structural reasons.

So why not just say that word meanings, or semantic concepts, are what changes when oncologists propose radically different ways of categorizing tumours, or the Obesity Society announces that obese people are diseased? As I argued above, first because if one takes the ontology of semantic concepts seriously, they are abstract entities, and need to be internalized to do some explanatory work in thought

[14] For a related response to the externalist problem, see (Chalmers, 2020).

and action; second, because of the externalist intuition that expert's conceptions—as said above—even the more recent and refined ones, about complex biological conditions such as diseases may just be false, and still count as our best effort at grasping what is true of them.

As conceptions are not currently very popular in the philosophical debate, let me point out in conclusion that my proposal is not as revisionary as it may seem, because of its limited scope. I suspect that the distinction between concepts and conceptions is particularly useful when we want to understand and change the dynamics of scientific discourse, for two reasons. First, because the externalist intuition, that concepts are of kinds of things out there that different people relate with and aim at knowing better, with the possibility of massive error, is arguably more compelling when we talk about cancers or mushrooms, than when we talk about, say, cultural appropriation or privacy. Second, because the concept-conception framework works well at explaining different levels of expertise on the same topic, and science is typically organized hierarchically when it comes to knowledge. Prima facie, however, it is not plausible that the rules of the game of medical research and healthcare management are the same as those of other human activities.

The thought here is that practices, knowledge hierarchies, relations of epistemic trust and deference, and metasemantic rules for change that hold within medicine (or maybe, other specialized sciences) need not be generalizable to other communities. I would rather speculate that in order to ameliorate teenagers' concept of obesity, so that it becomes less value-laden and discriminatory, a top-down operation on explicit definitions or constitutive principles would be pointless—where are such definitions supposed to be, in social media posts? Who's the authority? How are the boundaries of the community itself fixed? If this suspicion is well-founded, then it might be that different metasemantic frameworks can be chosen by conceptual engineers, for different domains and projects—for politics, morality, scientific contexts, or philosophy. Possibly in some of those, conceptions would be explanatorily idle. In fact, in the previous paragraphs I have characterized semantic concepts, psychological concepts and conceptions as constructs, meaning that what counts primarily is their explanatory function within a theoretical project. Taking my suspicion at face value, moreover, means opting for a radical pluralism about the objects of conceptual engineering (Belleri, 2021; Isaac & Koch, 2021; Isaac et al., 2022; Sawyer, 2021). My pluralistic attitude extends to the prospects of success of conceptual engineering projects. Speculatively, epistemically hierarchical communities with tight links between members, and explicit statements and discussions on basic conceptual issues, such as that of medical research, look like an ideal field for conceptual engineering. Experts' conceptions are, so to speak, open to view, and so are epistemic rules of argumentation and deference. I am not in a position to project this optimism to other domains, though. My considerations about pluralism, however, stand in need of a more thorough discussion, involving cases of non-medical concepts.

3.6 Conclusion

In this chapter, I illustrated two cases of ameliorative conceptual change in medicine involving cancer and obesity. In both cases, the amelioration project is approached with practical and ethical goals in view—at the bottom line, with the goal of doing less harm to patients, and having a better quality of healthcare. In the case of cancer, I showed how classifying, and demarcating involves choosing between different evidence-based criteria in a way that cannot be itself evidence-based, but rather value-based. In the case of obesity, I described the proposal of classifying obesity as a disease, aimed at bringing advantages to obese people in terms of more research on their condition, better care, and less stigma. My point in both cases was not to adjudicate whether the proposals are really conducive to amelioration or not. Rather, I wanted to show that medical research and healthcare organization are active fields of conceptual engineering.

In the remaining part of the chapter, I proposed that the most adequate metasemantic framework in the case of conceptual changes in medical research includes the notion of conception, and conceptual engineering in such area is better described as directed to expert conceptions. So, all my previous talk of concepts in the chapter ought to be replaced with conception talk. Expert's conceptions are incessantly rediscussed and reshaped, not only in the light of new evidence, but also and maybe primarily in the light of practical and ethical preferences and values. I argued that experts' conceptions are the preferred and a relatively easy target for conceptual engineering in the medical domain, in that they are public and explicit—therefore addressing the implementation problem for conceptual engineering. Medical and healthcare research, however, is a peculiarly structured activity, and different human practices and communities may call for different strategies for ameliorating our representational devices. This latter consideration opens up the possibility of a plurality of frameworks for conceptual engineering (Isaac & Koch, 2022).

This chapter is intended as the first step in a not-much-explored territory for philosophers of medicine interested in conceptual engineering. The metasemantic framework I proposed can be objected to, but I hope I pointed to a recognizable area of problems and work to be done. First, we need to investigate the details of how medical concepts or conceptions are formed and changed in various cases (a task which is close to the social epistemology of science and medicine). Here, a reflection on the non-epistemic basis of medical knowledge is in place: what is the role of values in such decisions? Which and whose values are involved? Starting from such investigations, then, there is room for a normative project, that of individuating the principles according to which such changes should be accomplished in general—including norms and criteria about inclusion and public participation.[15]

[15] My thanks are due to the Editors of this collection, for very useful comments to the first version of this Chapter, who made me rethink carefully about my metasemantic framework, and to Stefano Fanti for invaluable advice on all things medical.

References

Allison, D. B., Downey, M., Atkinson, R. L., Billington, C. J., Bray, G. A., Eckel, R. H., Finkelstein, E. A., Jensen, M. D., & Tremblay, A. (2008). Obesity as a disease: A white paper on evidence and arguments commissioned by the council of the obesity society. *Obesity, 16*(6), 1161–1177.

Amoretti, M. C., & Lalumera, E. (2020). The concept of disease in the time of COVID-19. *Theoretical Medicine and Bioethics, 41*(5), 203–221.

Amoretti, M. C., & Lalumera, E. (2021). *Wherein is the concept of disease normative?* Medicine, Health Care and Philosophy online first. https://doi.org/10.1007/s11019-021-10048-x

Belleri, D. (2021). On pluralism and conceptual engineering: Introduction and overview. *Inquiry, 0*(0), 1–19.

Bayer, R. (1987). *Homosexuality and American Psychiatry: The Politics of Diagnosis*. Princeton University Press.

Boorse, C. (1977). Health as a theoretical concept. *Philosophy of Science, 44*(4), 542–573.

Boorse, C. (1997). A rebuttal on health. In J. M. Humber & R. F. Almeder (Eds.), *What is disease?* (pp. 1–134). Humana Press.

Bosman, F. T., Carneiro, F., & Hruban, R. H. (2010) *WHO classification of tumours of the digestive system: WHO classification of tumours, Volume 3*. World Health Organization.

Broadbent, A. (2019). *Philosophy of medicine*. Oxford University Press.

Bueter, A. (2019). Epistemic injustice and psychiatric classification. *Philosophy of Science, 86*, 1064–1074. https://www.cambridge.org/core/journals/philosophy-of-science/article/abs/epistemic-injustice-and-psychiatric-classification/7E6F4986A24372AD2FA182DB4D071D57

Burge, T. (1979). Individualism and the mental. *Midwest Studies in Philosophy, 4*, 73–121.

Burgess, A., & Plunkett, D. (2020). On the relation between conceptual engineering and conceptual ethics. *Ratio, 33*, 281–294. https://onlinelibrary.wiley.com/doi/abs/10.1111/rati.12265

Callard, F., & Perego, E. (2021). How and why patients made long COVID. *Social Science & Medicine, 268*, 113426. https://www.sciencedirect.com/science/article/pii/S0277953620306456

Cappelen, H. (2018). *Fixing language: An essay on conceptual engineering*. Oxford University Press.

Cappelen, H., & Plunkett, D. (2020). A guided tour of conceptual engineering and conceptual ethics. In A. Burgess, H. Cappelen & D. Plunkett (Eds.), *Conceptual engineering and conceptual ethics* (pp. 1–26). New York.

Carter, H. B., Partin, A. W., Walsh, P. C., Trock, B. J., Veltri, R. W., Nelson, W. G., Coffey, D. S., Singer, E. A., & Epstein, J. I. (2012). Gleason score 6 adenocarcinoma: Should it be labeled as cancer? *Journal of Clinical Oncology, 30*(35), 4294–4296.

Chalmers, D. J. (2020). What is conceptual engineering and what should it be? *Inquiry*, 1–18.

Cooper, R. (2014). *Diagnosing the diagnostic and statistical manual of mental disorders* (p. xvi, 79). Karnac Books.

Corrigan, P. W., & Bink, A. B. (2016). The stigma of mental illness. In *Encyclopedia of mental health* (pp. 230–234). Elsevier. https://doi.org/10.1016/B978-0-12-397045-9.00170-1

de Vries, J. (2007). The obesity epidemic: Medical and ethical considerations. *Science and Engineering Ethics, 13*(1), 55–67.

Deutsch, M. (2020). Speaker's reference, stipulation, and a dilemma for conceptual engineers. *Philosophical Studies, 177*(12), 3935–3957. https://doi.org/10.1007/s11098-020-01416-z

Deutsch, M. (2021). Still the same dilemma for conceptual engineers: Reply to Koch. *Philosophical Studies, 178*(11), 3659–3670. https://doi.org/10.1007/s11098-021-01619-y

Douglas, H. (2009). *Science, policy, and the value-free ideal*. University of Pittsburgh Press.

Dresser, R. (2012). *Malignant: Medical ethicists confront cancer*. Oxford University Press.

Dutilh-Novaes, C. (2018). Carnapian explication and ameliorative analysis: A systematic comparison. *Synthese*, 1–24.

Epstein, J. I., Egevad, L., Amin, M. B., Delahunt, B., Srigley, J. R., & Humphrey, P. A. (2016). The 2014 International Society of Urological Pathology (ISUP) consensus conference on Gleason grading of prostatic carcinoma. *The American Journal of Surgical Pathology, 40*(2), 244–252.
Epstein, J. I. (2022). Is Grade Group 1 (Gleason Score $3 + 3 = 6$) Adenocarcinoma of the Prostate Really Cancer?, *Current Opinion in Urology*, 32, 91–95.
Esserman, L. J., Thompson, I. M., Reid, B., Nelson, P., Ransohoff, D. F., Welch, H. G., Hwang, S., Berry, D. A., Kinzler, K. W., Black, W. C., Bissell, M., Parnes, H., & Srivastava, S. (2014). Addressing overdiagnosis and overtreatment in cancer: A prescription for change. *The Lancet Oncology, 15*(6), e234–e242.
Evans, G., & Altham, J. E. J. (1973). The causal theory of names. *Proceedings of the Aristotelian Society, Supplementary Volumes, 47*, 187–225. https://www.jstor.org/stable/4106912
Fanti, S., Oyen, W., & Lalumera, E. (2019). Consensus procedures in oncological imaging: The case of prostate cancer. *Cancers, 11*(11), 1788.
Fischer, E. (2020). Conceptual control: On the feasibility of conceptual engineering. *Inquiry, 68*(9), 3043–3071. https://doi.org/10.1080/0020174X.2020.1773309
Flocke, V. (2021). How to engineer a concept. *Philosophical Studies, 178*, 3069–3083. https://doi.org/10.1007/s11098-020-01570-4
Fuller, J. (2018). What are chronic diseases? *Synthese, 195*(7), 3197–3220.
Goffman, E. (2009). *Stigma: Notes on the Management of Spoiled Identity*. Simon and Schuster.
Gutin, I. (2021). Body mass index is just a number: Conflating riskiness and unhealthiness in discourse on body size. *Sociology of Health & Illness, 43*, 1437–1453. http://onlinelibrary.wiley.com/doi/abs/10.1111/1467-9566.13309
Häuser, W., & Fitzcharles, M.-A. (2018). Facts and myths pertaining to fibromyalgia. *Dialogues in Clinical Neuroscience, 20*(1), 53–62.
Hofmann, B. (2002). On the triad disease, illness and sickness. *The Journal of Medicine and Philosophy: A Forum for Bioethics and Philosophy of Medicine, 27*(6), 651–673.
Hofmann, B. (2014). Diagnosing overdiagnosis: Conceptual challenges and suggested solutions. *European Journal of Epidemiology, 29*(9), 599–604.
Hofmann, B. (2016). Obesity as a socially defined disease: Philosophical considerations and implications for policy and care. *Health Care Analysis, 24*(1), 86–100.
Isaac, M. G. (2020). How to conceptually engineer conceptual engineering? *Inquiry, 67*(1), 199–214.
Isaac, M. G. (2021). Post-truth conceptual engineering. *Inquiry*, 1–16. https://dx.doi.org/10.1080/0020174X.2021.1887758
Isaac, M. G., & Koch, S. (2022). Foundational issues in conceptual engineering: introduction and overview. *Inquiry, 0*, 1–9. https://doi.org/10.1080/0020174X.2022.2028230
Jastreboff, A. M., Kotz, C. M., Kahan, S., Kelly, A. S., & Heymsfield, S. B. (2019). Obesity as a disease: The obesity society 2018 position statement. *Obesity, 27*(1), 7–9.
Jorem, S. (2020). Conceptual engineering and the implementation problem. *Inquiry, 64*(1–2), 186–211. https://doi.org/10.1080/0020174X.2020.1809514
Koch, S. (2021a). Engineering what? On concepts in conceptual engineering. *Synthese, 199*, 1955–1975. https://doi.org/10.1007/s11229-020-02868-w
Koch, S. (2021b). The externalist challenge to conceptual engineering. *Synthese, 198*, 327–348. https://doi.org/10.1007/s11229-018-02007-6
Kripke, S. A. (1972). Naming and necessity. In D. Davidson & G. Harman (Eds.), *Semantics of natural language* (pp. 253–355). Dordrecht: Springer Netherlands. https://doi.org/10.1007/978-94-010-2557-7_9
Lalumera, E. (2010). Concepts are a functional kind. *Behavioral and Brain Sciences, 33*, 217–218. https://www.proquest.com/docview/527775419/abstract/1AA6C9B8267C4A64PQ/1
Lalumera, E. (2014). On the explanatory value of the concept-conception distinction. *Rivista Italiana di Filosofia del Linguaggio, 8*(3), 73–81.
Laurence, S., & Margolis, E. (2003). Concepts and conceptual analysis. *Philosophy and Phenomenological Research, 67*, 253–82. https://onlinelibrary.wiley.com/doi/abs/10.1111/j.1933-1592.2003.tb00290.x

Lemoine, M. (2013). Defining disease beyond conceptual analysis: An analysis of conceptual analysis in philosophy of medicine. *Theoretical Medicine and Bioethics, 34*(4), 309–325.
Longino, H. E. (1996). Cognitive and non-cognitive values in science: Rethinking the dichotomy. In L. H. Nelson & J. Nelson (Eds.), *Feminism, science, and the philosophy of science* (pp. 39–58). Springer.
Lysdahl, K. B., & Hofmann, B. (2020). Overutilization of imaging tests and healthcare fairness. In E. In Lalumera & S. Fanti (Eds.), *Philosophy of advanced medical imaging* (pp. 99–111). Springer.
Machery, E. (2017). *Philosophy within its proper bounds*. Oxford University Press.
Müller, M. J., & Geisler, C. (2017). Defining obesity as a disease. *European Journal of Clinical Nutrition, 71*, 1256–1258. https://www.nature.com/articles/ejcn2017155
Peacocke, C. (1992). *A study of concepts* (pp. xiii, 266). Cambridge, MA: The MIT Press.
Peacocke, C. (1998). Implicit conceptions, understanding and rationality. *Philosophical Issues, 9*, 43–88. https://www.jstor.org/stable/1522959
Pinder, M. (2021). Conceptual engineering, metasemantic externalism and speaker-meaning. *Mind, 130*, 141–163. https://doi.org/10.1093/mind/fzz069
Plutynski, A. (2018). *Explaining cancer: Finding order in disorder*. Oxford University Press.
Pollock, J. (2021). Content internalism and conceptual engineering. *Synthese, 198*, 11587–11605. https://doi.org/10.1007/s11229-020-02815-9
Purysko, A. S., Magi-Galluzzi, C., Mian, O. Y., Sittenfeld, S., Davicioni, E., du Plessis, M., Buerki, C., Bullen, J., Li, L., Madabhushi, A., Stephenson, A., & Klein, E. A. (2019). Correlation between MRI phenotypes and a genomic classifier of prostate cancer: Preliminary findings. *European Radiology, 29*(9), 4861–4870.
Putnam, H. (1975). The Meaning of "Meaning". *Minnesota Studies in the Philosophy of Science, 7*. https://philpapers.org/rec/PUTTMO
Richiardi, L., Barone-Adesi, F., & Pearce, N. (2017). Cancer subtypes in aetiological research. *European Journal of Epidemiology, 32*(5), 353–361.
Riggs, J. (2019). Conceptual engineers shouldn't worry about semantic externalism. *Inquiry, 0*, 1–22. https://doi.org/10.1080/0020174X.2019.1675534
Rudner, R. (1953). The scientist qua scientist makes value judgments. *Philosophy of Science, 20*(1), 1–6.
Russo, A., et al. (2021). *Practical medical oncology textbook*. Springer International Publishing.
Salto-Tellez, M., & Cree, I. A. (2019). Cancer taxonomy: Pathology beyond pathology. *European Journal of Cancer, 115*, 57–60.
Sanders, R. (2019). The color of fat: racializing obesity, recuperating whiteness, and reproducing injustice. *Politics, Groups, and Identities, 7*, 287–304. https://doi.org/10.1080/21565503.2017.1354039
Sawyer, S. (2018). The importance of concepts. *Proceedings of the Aristotelian Society, 118*(2), 127–147.
Sawyer, S. (2020). Truth and objectivity in conceptual engineering. *Inquiry, 63*(9–10), 1001–1022.
Sawyer, S. (2021). Concept pluralism in conceptual engineering. *Inquiry, 0*(0), 1–26.
Scharp, K. (2013). *Replacing Truth*. OUP Oxford.
Schramme, T. (2017). Goals of medicine. In T. Schramme & S. Edwards (Eds.), *Handbook of the philosophy of medicine* (pp. 121–128). Springer.
Schwartz, P. H. (2007). Defining dysfunction: Natural selection, design, and drawing a line. *Philosophy of Science, 74*(3), 364–385.
Schwartz, P. H. (2014). Small tumors as risk factors not disease. *Philosophy of Science, 81*(5), 986–998. https://doi.org/10.1086/678280
Seale, C. (2001). Sporting cancer: Struggle language in news reports of people with cancer. *Sociology of Health & Illness, 23*(3), 308–329.
Sholl, J. (2020). Health in philosophy: Definitions abound but a theory awaits. In J. Sholl, & S. I. S. Rattan (A c.Di) (Eds.), *Explaining health across the sciences* (pp. 79–95). Springer International Publishing.
Simion, M. (2017). The 'should' in conceptual engineering. *Inquiry*, 1–15.

Solomon, M. (2016). Expert consensus. In J. Simon & H. Kincaid (Eds.), *The Routledge Companion to Philosophy of Medicine* (pp. 367–374). Routledge.

Stegenga, J. (2016). Three criteria for consensus conferences. *Foundations of Science, 21*(1), 35–49.

Stegana, J. (2018). *Medical Nihilism*. Oxford University Press.

Tekin, Ş. (2022). Participatory interactive objectivity in psychiatry. *Philosophy of Science, 89*, 1166–1175. https://www.cambridge.org/core/journals/philosophy-of-science/article/participatory-interactive-objectivity-in-psychiatry/11F0F0A2DF37EEDADEFB7B8A4E275877#

Travis, W. D., et al. (2015). The 2015 World Health Organization Classification of Lung Tumors: Impact of Genetic, Clinical and Radiologic Advances Since the 2004 Classification. *Journal of Thorac Oncology, 10*(9), 1243–1260. https://doi.org/10.1097/JTO.0000000000000630

Tsang, J. Y. S., & Tse, G. M. (2020). Molecular classification of breast cancer. *Advances in Anatomic Pathology, 27*(1), 27–35.

Varghese, A. (2015). Chemotherapy for stage II colon cancer. *Clinics in Colon and Rectal Surgery, 28*(4), 256–261.

Wittgenstein, L. (2010). *Philosophical investigations*. John Wiley & Sons.

Chapter 4
Rethinking the Ocean: Conceptual Engineering for the Next 10,000 Years

Roberto Casati, Sacha Bourgeois-Gironde, Emilie Flamme, Alberto Gallace, Quentin Hiernaux, Stefano Malatesta, Marcella Schmidt di Friedberg, Enrico Squarcina, Colomban de Vargas, and Eva Wanek

Abstract The ocean constitutes an interesting case study for conceptual negotiation and reconceptualization, in view of its otherness, of the current pressure exerted by anthropic activities on many aspects of oceanic status and life, of new, emerging knowledge about its physical and biological nature, and in particular in view of the need to look for and plan actions on a long time span, measured not in decades, but in centuries and millennia. The present contribution analyzes existing conceptualizations of the ocean that in the current context lean towards misconceptions, and in its programmatic part proposes reconceptualizations that can eventually inform public policies.

Keywords Conceptual engineering · Conceptual negotiation · Ocean · Anthropocene · Climate change · Reconceptualization of the ocean

---

R. Casati (✉) · E. Flamme · Q. Hiernaux · E. Wanek
Institut Jean Nicod, UMR 8129, ENS-EHESS-CNRS, DEC, Université PSL, Paris, France
e-mail: roberto.casati@ens.fr

S. Bourgeois-Gironde
Université Paris 2 Panthéon-Assas - Centre for Law and Economics, Paris, France

Università Milano Bicocca, Milan, Italy

A. Gallace · S. Malatesta · M. S. di Friedberg
Università Milano Bicocca, Milan, Italy

E. Squarcina
Université Paris 2 Panthéon-Assas - Centre for Law and Economics, Paris, France

C. de Vargas
Sorbonne Université, CNRS, Station Biologique de Roscoff, UMR7144, ECOMAP & Plankton Planet, Roscoff, France

Research Federation for the study of Global Ocean Systems Ecology and Evolution, FR2022/Tara GOSEE, Paris, France

M. G. Isaac et al. (eds.), *New Perspectives on Conceptual Engineering - Volume 3*, Synthese Library 515, https://doi.org/10.1007/978-3-031-98535-5_4

4.1 Introduction

The main body of research about notions of the ocean in the last years, mostly in the wake of Steinberg's seminal *The Social Construction of the Ocean* (Steinberg, 2001) has been devoted to highlighting powerful historical, social political and (at times) cognitive biases in the current representations of the ocean, of its non-human inhabitants, and of relevant actors and stakeholders. Little appears to have made its way in the literature in terms of proposing *alternative representations*, and very little in terms of integrating biological and physical knowledge into the scheme. However, biases in ocean representation have been alleged to inflect policies and behaviors that—given the current Anthropocene crisis—can be highly counterproductive (Rock et al., 2020). The present chapter, drawing on the theory of conceptual negotiation (Casati, 2011), one of the mechanisms of conceptual engineering, proposes several *ad hoc* reconceptualizations of notions used in describing the ocean and the animate or inanimate entities living in it.

The ocean constitutes an interesting case study for conceptual negotiation and reconceptualization, in view of its otherness, of the current pressure exerted by anthropic activities on many aspects of oceanic status and life, of emerging knowledge about its physical and biological nature, and in particular in view of the need to look for and plan ahead actions on a long time span, measured not in decades, but in centuries and millennia. Focusing specifically on the ocean, the UNESCO's "One Planet, One Ocean" initiative intimates that "we need to change the way we *think* and act" (UNESCO, 2013). The present contribution analyzes existing conceptualizations of the ocean that in the current context lean towards misconceptions and proposes reconceptualizations that can eventually inform public policies.

Some facts contextualize the current project. The sea is far from the eyes of a large share of the world population; this "invisibility" contributes to negligent policies, in particular around concerns of extraction and pollution. The ocean is a fundamental resource for life on the planet and for humans in particular as recognized, if belatedly, in the preamble to the 2015 Paris Agreement, which affirms "the importance of ensuring the integrity of all ecosystems, including oceans", reinstated in the 2021 Glasgow Climate Pact, where it is noted "the importance of ensuring the integrity of all ecosystems, including in forests, the ocean and the cryosphere". A liter of seawater contains between 10 billion and 100 billion microorganisms, "forming a continuous global ecosystem that generates approximately half of planetary oxygen, sustains the large majority of marine life, and regulates atmospheric CO2 and climate" (de Vargas et al., 2020; Falkowski, 2012). The ocean is massively used as a resource for human food (Holm & Macdiarmid, 2017), where it has been driven to the brink of depletion, in particular in the age of mechanized, industrial fishing (Fagan 2017). Almost 90% of world shipping goes through maritime routes (Tasse & Alex, 2020): the ocean is a cultural, merchant, and social connector, and the main vector of globalization. The ocean bottoms are resourceful, rich in oil and minerals, and their exploitation, although contested, is well under way. At the same time, the ocean is the major dump of human-produced chemicals and of human debris in general (Tani, 2020), including nuclear waste (Abella et al., 2019) and even decommissioned satellites (De Lucia & Iavicoli, 2018). Most importantly, the ocean has cushioned about 30% of the CO2 generated by anthropogenic activities, 94% of all the surplus heat (1% of the heat is absorbed by the atmosphere, the rest by continents and continental ice melting) (Speich, 2019). This means that the ocean has accumulated enormous internal energy, has increased in volume, has become more acidic and contains less oxygen. Finally, seas connect, but are also conceptualized as boundaries, and are construed as such in regulating people's movements or in disputes about that regulation (Maccanico, 2019).

If we target the preservation of the ocean as one of the major issues in climate and societal change, and as the one where there are large margins for effective societal intervention, we can locate two main issues in (a) the well-documented difficulty of inducing massive behavioral change in the absence of well-defined and binding legal frameworks, and (b) in the insufficient salience and inadequate conceptual representations of the ocean, which hamper the delineation of effective measures and policies.

Research in the behavioral sciences acknowledges that there exists a wide spectrum of possibilities to modulate behavior, from purely informational approaches to nudges, from campaigns to educational interventions, from object design to the implementation of incentives/disincentives, to institutional design (Benabou & Tirole, 2011). However, the sheer scale of the change needed in the Anthropocene, where human action has consequences measured in geological time-spans, requires a fresh approach; in particular, research shows that, in general, *massive* behavioral change is extremely hard (Nisa et al., 2019), that law-framed measures tend to be more effective than campaigns, but, at the same time, that measures require compliance and that the latter is not to be taken for granted (Millock & Nauges, 2010). Finally, it is important to underline that behavioral change is standardly understood as *individual* behavioral change, thereby supporting non-adaptive policies that, on the one hand, are overloading individuals with impossible-to-match responsibilities and are conducive to climate anxiety (Clayton, 2020) and that, on the other hand, risk generating only marginal change.

All this motivates focusing on a framework that would be hospitable to the articulation of innovative policies, that could be readily accepted by citizens. Policies, however, can only be as good as their characterization of the problem they address. And here is where conceptual negotiation intervenes, in a process that assesses current concepts and suggests way to improve them. (Mis)conceptions about the ocean (e.g., as boundless, as a hyper-resource (Casati, 2022a), as alien and thereby "distant" or unrelated to our lives, as conceptually tributary to the land, among others) run deep in public and media discourse, public policies, business practices, education, the law, and individual and collective behavior. Policies should, at a minimum, incorporate more deeply a correct ontological characterization of the natural entity with respect to which they formulate rules and regulations. Characterizing the ocean as a *resource* expresses an ontology associated with industrial interests.[1] Characterizing it as a *territory* is a projection of political interests. And in neither of these categorizations, the intrinsic, *living* nature of the ocean is accounted for. A *conceptual resistance* quickly arises by objecting that an inert substance has no subjective interests. This is what contemporary legal innovations are designed to bypass, by stipulating some intrinsic interests to that (supposedly) inert substance and, for instance, by associating personal rights to those interests (Stone, 2010). The ad-hoc creation of oceanic legal persons is a cognitive reconfiguration of oceanic entities, placing nature at the core of legal conceptualization, and on a practical level, the interests of both human and natural entities on a par.[2]

And indeed, the ocean is at present legally conceptualized in different, a times mutually inconsistent ways: as a *resource*, as a set of *resources*, or as a *means to circulate resources* and it is further conceived in terms of a *vast divisible territory* over which different levels of national sovereignty and international cooperation can be exerted (Nordquist et al., 2018). The focus on legal regulations of resources extraction and on territorial rights stem from two main concerns driving the standard legal conceptualization of oceans: wealth and power, economics, and politics. Both currently fail to solve the most outsized, unprecedented externality human activities have ever produced: global warming. Alternate conceptual representations proposed in the biological and geophysical literature include the notion of *unity and indivisibility of the ocean*, both spatially and in terms of internal dynamics (Jacques et al., 2020). These unity-focused reconceptualizations create a framework that is hospitable to more sensitive policies and legal instruments, such as the management of commons and the attribution of legal personality to the ocean.

The reconceptualization move is acknowledged in the literature. Discussing the evolution of commons, Bravo and Moor (2008) write that "... many commons have changed: new functions arose, new users, and new institutional layers emerged to manage them. Even for some very traditional commons, like the alpine pastures, the focus of today's management is no longer only a matter of producing fodder and at the same time keeping the cattle numbers below the carrying capacity of the meadow, but also of a more general concern of landscape and hydro-geologic stability maintenance. Similarly, forests are no longer considered only as a source of timber and other products but are increasingly regarded as carbon sinks, as water-capture and water-conservation tools, and as leisure areas."

---

[1] Consider the legal and political negotiations about Biodiversity Beyond National Jurisdiction.

[2] Legal personhood is in principle detached from any metaphysical implications concerning the fact that the bearer of this juristic fiction is a living, sentient, conscious being or not. As a fiction it says nothing about biological facts. However, as Salmond (1893) formulated it, legal personality brings something to existence which otherwise is not. Thus, subjectivity has also to be understood in terms of this legal existence—a subject came to existence—not as if we actually recognized, metaphysically, that oceanic entities, living or not, were endowed with a phenomenological ego - something which belongs to another sort of discussion, and can incidentally or substantially inform the law, but from which legal creation—seen positively—is in principle independent.

This chapter enlarges the strategy: it targets the identification of a sizable number of core (mis)conceptions and proposes conceptual negotiations to assess them and implement adequate and effective conceptual replacement, i.e., reconceptualizations. Candidate misconceptions will be prefixed by a star ‘*’, dubious conceptions by a dubitative mark ‘⸮’. Section 4.2 provides a general conceptual negotiation framework for reconceptualizations, briefly engaging with the literature on conceptual engineering. Section 4.3 presents a list of specific oceanic challenges and reconceptualizations. This is the core of the chapter, providing a spectrum of problems that the proposed reconceptualizations address. Section 4.4 enlists limit cases that can be used as a textbed for reconceptualization programs. Section 4.5 hints at future work.

4.2 Conceptual Negotiation and Misconceptions Management

4.2.1 Conceptual Negotiation

According to the framework of conceptual negotiation (Casati, 2011), new concepts are created, tested, and circulated in situations in which existing concepts that permeate public discourse and individual and collective decision-making crack under the pressure from new societal demands or scientific leaps. The negotiations do not necessarily happen in academic contexts. They can occur in court, as was the case of *Brancusi vs United States* 1928 trial (Hartshorne, 1986) which reconceptualized *art* as a *non-utilitarian object* per customs requirements. Negotiations have also occurred in constitutional assemblies, as in the case of Italy’s transition from monarchy to republic and the reconceptualization of individuals from *subjects* to *citizens*, and in the scientific arena, as in the case of paradigm shifts. These scientific shifts include the move from the Ptolemaic to the Copernican cosmology implying a deep reconceptualization of *planets*, as well as the 2006 demotion of Pluto, no longer conceptualized as a planet by the International Astronomical Union (IUA, 2006) when the concept of a planet was no longer able to sustain the pressure generated by the discovery of a number of bodies of comparable size in the Kuiper Belt.

The new concepts can be based on a better understanding of underlying factual elements, on scientific consensus (Drouet et al., 2021), but also on societal demands and on shifting borders of understandability and social accountability (Haslanger, 2000). Conceptual negotiations accompany the transitions, along paths that are well travelled in academic philosophy (e.g., use of thought experiments, focus on limit cases, reworking of definitions/characterizations/narratives along incremental parametric changes, taking advantage of analogies with well understood cases, running analyses of the tradeoffs among proposed conceptual changes, inducing perspectival change.) Arguably, new concepts should have positive effects—facilitating agreement, providing new perspectives and repertories that unlock action, preventing negative actions, de-invisibilizing important issues or stakeholders, operationalizing the up to then unthinkable (Stone, 2010). A very simple example of an action-unlocking negotiation is how one-way streets in some French towns have been reconceptualized as *two-way streets for cyclists*. Remaining in the domain of surface transportation, roads are conceptualized as *connectors*, while their physical presence in a territory turns them into *barriers* to the movement of living entities (Blandin, 2020). Negotiation processes offer a secluded and protected model of conceptual change whose consequences are imagined and contribute to a mutual understanding between parties—as any rational negotiation ought to (Fisher et al., 1997). Negotiations broaden the spectrum of alternatives through agreement on shared normative criteria.

4.2.2 Misconceptions and Possible Actions About Them

Misconceptions must first be identified and deconstructed. Sometimes they disguise genuinely "fake concepts", i.e. concepts that purport to describe a segment of reality but do so in misleading ways, characterized as they are by a mix of empirical inadequacy, conceptual murkiness, iffy motivations and hidden agendas, c.f., (Krugman, 2020) "Zombie theories" in economics (disproven theories that keep resurfacing in public debate). As an instance of fake concepts, **digital natives, *multitasking, *access to knowledge* have been heralded as main conceptual vectors of the **digital migration* (Casati, 2013). Their deconstruction opens the way to fitter reconceptualizations, such as ˁ*digital subjects, task switching, access to information, digital colonialism,* respectively. A neat example of a fake concept is the notion of **natural gas*, where the modifier "natural" suggests some kind of eco-friendly substance; gas labeled as "natural" should be reconceptualized without hesitation as *fossil gas*.

The subsequent reengineering of concepts does not follow a single path. Conceptual engineers can avail themselves of several strategies, viz. concept introduction, replacement, revision, and elimination (Isaac et al., 2022). Each strategy could then use different resources. For instance, concept revision could draw from.

- *genus shift* (e.g., *sea rescue* when reconceptualized as **crime* (Maccanico, 2019), thus no longer a species of the genus *moral imperative*).
- *extension* (be it conservative or not: the **inanimate sea* and the inanimate Earth can be reconceptualized as *living beings,* e.g., per the Gaia hypothesis (Lovelock, 1990) or as a *legal persona*,
- *unification* (when *sea* and *ocean* are no longer seen as genera of the same species, or as related by genus-species relations),
- *analogy mining* (e.g., plankton activity rethought as a *service* or even as *labor*).

Furthermore, the reconceptualization strategy is not necessarily stable; no general rules are available, and in some cases (**witch*, **phlogiston*, **digital native*) outright concept *elimination* may be the best or even the only viable option (e.g., **Mediterranean climate* does not appear to be a scientifically supported notion).

It should be noted that new concepts are not automatically *better* concepts, as innovation is not *per se* progress. Whether it is or not depends of course on contextual and evaluative factors. When *sea rescue* is reconceptualized as *crime*, some may find the new determinable fitter, others less fit. Pending a final normative statement on the quality of the new concept, what minimally matters is the quality of the negotiation process, or even more minimally, that a process exists. This is particularly salient in in the Whanganui case (Bourgeois-Gironde, 2020), whereby a natural entity—a river—has acquired *legal personhood*; the process cares that the conversation is not just hostage of marketing strategies, of top-down conceptual imposition, or of agenda setting. The issue has a formal aspect: a new player arrives in a juridical negotiation game (on the use of the natural entity and the fact of defining the river as a *player* and no longer as a **resource* changes the equilibrium of this game, or even transforms a unilaterally human decision into a *de facto* game. We can also consider that agreements such as the one involving the Whanganui River are simply the result of a very long process of negotiation and that the idea is that negotiation continues within the Te Awa Tupua/Whanganui River Claims Settlement. What this agreement implements is a continued process framed in a new form of negotiation game.

Another lever of conceptual change is re-perspectivalization, a spectacular example of which comes from the history of astronomy. In "Kepler's move" (Anderson, 1969), Copernican Earth's spin is acknowledged as cognitively unfriendly, given that visual perception invariably presents the Earth as **still* and celestial bodies in **slow revolution* around it. An imaginary change of perspective can assist: Kepler asked his readers to put themselves in the shoes of an astronomer on the visible side of the Moon; from there they would *see* that the Earth *spins*, about 28 times during one of their lunar days. Importantly, the thought experiment does not prove that the Earth spins. However, it makes the concept of a spinning Earth fluid and more acceptable. This would be a case of extreme perspectivalization: physically go there and look! Media and art can (Dupont, 2017), in some measure, obviate the difficulty of taking people out on spaceships or on boats on the high seas to address cognitive distance—we know the powerful effect similar reperspectivalization moves had on public opinion at the beginning of space travel, when pictures of the Earth from outer space became available (Boyle et al., 2018).

Finally, reviewing existing misconceptions and assembling new perspectives and reconceptualizations on the ocean must be based on evidence from a large panel of sources: climate science, geography, biology, history, philosophy, the law, economics, and know-how and viewpoints of ocean-sensitive and special-need communities. Of course, this adumbrates a whole research program and here it will be only possible to superficially broach the topic.

4.3 Getting the Work Done: Oceanic Concept Challenges

4.3.1 Dimensions of Otherness

The challenge is not small, in particular as the sea has been perceived as *other* in almost every cultural sphere, even by peoples who have lived at sea and of the sea. For instance, the Western philosophers studied by Scholtz (2016) did not deploy a systematic approach to the sea; it is too different, relegated to the background, almost a topic to be avoided; when it is narrated, its contrasting aspects with the land are highlighted, and notably its inhumanity. The Flood in the Bible makes the Earth, albeit temporarily, a nightmarish waterworld, and it is telling that John in the Apocalypse invokes a paradise awkwardly freed from the sea: "I saw a new heaven and a new earth, because the heaven and the earth of before had disappeared and the sea was no more" (Connery, 2006).

What are the dimensions of this otherness? When we try to assess our complex representations of the ocean we can look at the contrast class of our adaptive environment, the one our ancestors inhabited long enough to develop and sediment in our genetic memory a predilection that still endures for the things that comfort our desire for well-being as well as our artistic and poetic imagination: stable ground beneath our feet, shady woods, shelters, water sources, a landscape rich in information, scents, beaten paths, and stable, distant and clearly visible landmarks such as mountains or isolated trees. The sea is the *dual image of* this idyllic scene. The open sea is a human desert that offers no shelter from the sun, is without protecting shadows, has an indefinite and changing color, is made of a kind of water that cannot be drunk without being processed first, is in constant movement, is beaten by violent winds that raise large and powerful waves, is run by strong currents, hides perilous and invisible abysses, and is closed in every direction by a uniform and uninformative horizon. It presents us with a paradoxical surface on which we cannot walk, but which, being free of obstacles, affords an irresistible invitation to journey. If deserts of sand and rock are *hostile*, the sea is *alien*. The difficulty of policing the ocean and of enforcing regulations further creates a special case for any legal approaches to this otherness.

Liquidity and size turn the sea into a straightforward challenge to the conceptual abilities of the human mind. Some of the difficulties are general and depend on conflicting naive conceptualizations about liquid entities (Hayes, 1985). Other difficulties may even be not specific to the water element: even land delimitations, in spite of numerous attempts to ground political geography on physical geography, and in spite of the existence of recognized criteria for watersheds, suffer from a large share of arbitrariness. But the sea erases every boundary; and since all the seas of the world converse with all the other seas, the boundaries that history has traced in the sea are purely conceptual (Frege & Thiel, 1986) or derive their geometry from some aspect of the nearest coast. For instance, the boundaries of territorial waters or of Exclusive Economic Zones are parasitic on the geometry of coastlines.

We do not even know what deserves to be *called* "sea" or "ocean", with what right these names are assigned, what exactly they refer to. The names of the seas have many origins: they come from neighboring lands ("*Indian* Ocean" or "*Java* Sea"), from a local perspective ("*Inland Sea*" or "*North* Sea"), from a first encounter that generates a fortuitous impression ("*Pacific* Ocean"), by what one finds there ("*Sargasso* Sea"), by who is Eurocentrically believed to be its discoverer ("Strait of *Magellan*", whose original name was moreover "Strait of All Saints", because of the date of its first European navigation); and the water bodies corresponding to these names are largely arbitrary. And is the Caspian a sea or a big lake?

4.3.2 A Mid-Century Tension

In the last three centuries a large amount of knowledge about the ocean has been amassed (Jacques et al., 2020), but the science-inspired image is not devoid of internal tensions. In the span of 3 years, from 1951 to 1954, two books were published which epitomize the complex approach to the ocean in the middle of the twentieth century. The celebrated *The Sea Around Us* by Rachel Carson brings for the first time to the general public a conservationist look on the natural entities, foreshadows global warming, and warns about the risk of dumping nuclear waste in the ocean (Carson, 1951). The less known *The Inexhaustible Sea* by Francis Minot and Daniel Hawthorne, scientifically very well informed, puts all the knowledge available at the time at the service of the project to extract the maximum amount of protein from the sea to feed humanity (Hawthorne & Minot, 1954). Both works plead for the set-up of large-scale scientific programs to study the ocean, contribute to a de-invisibilization of the ocean, and invite to better take the ocean into account in public debates. But the tension between conservation and exploitation persists and runs through contemporary approaches to the ocean.[3]

4.3.3 Conflicting Conceptualizations

Existing conceptualizations may thus be inadequate or may be partial and enter in mutual conceptual conflict. The sea has been conceptualized both as an **inexhaustible resource* but also as a *limited resource,* whereby it is rational to protect it through territorialization; as an **alien space* (as recorded in a saying; "There are three sorts of people: those who are alive, those who are dead, and those who are at sea", (Lieutaud, 2014) but also as a much desired background of personal development and well-being (Nichols, 2015) or alternatively, as an entity to be

[3] Notably, Carson (1951) herself devotes several sections to what she considers the exciting possibility of extracting resources from the sea.

**superseded* (Connery, 2006) and **mentally canceled*; as a *connector*, whereby it must be free from enclosures (Grotius, 2012) so as to allow free movement on its surface, but also as a ⸮*frontier*, whereby it must be policed; as ⸮*purely material body of water* but also as a ⸮*living entity*, even as an ⸮*ancestor*, the latter characterization promoting an intriguing synergy between some ancient narratives of the origin of humans, and recent scientific, Darwinian theorizing (Hermitte, 2011).

A survey by Sugimoto et al. (2022), on perception and values attached to the Ocean on Okinawan islands, shows that resources provided by the sea are considered locally as *lavish gifts*: the ocean is bountiful, and that this perception of obtained sea-food as marine gifts, makes it difficult to cognitively accept the relevance of a purely quantitative nature of *economic services* the ocean and its inhabitants would provide. Services and gifts–in terms of the type of measures they involve–enter in mutual conflict. Another salient dimension tested through the Okinawa survey is the "fear and respect for nature" dimension. Typhoons, in particular, are feared, but they are given a positive valence by Okinawan islanders, as *purificatory events*, necessary for the good health of the ocean.

Ever deeper hesitations testify to the fact that concepts evolve (Vosniadou, 2008), and if differences are sometimes easy to locate, they turn out to be particularly hard to reconcile. The ocean can be thought of as a *space* (i.e., as per the 1982 Montego Bay Convention) but also as the *content* of that space; as a *physical system* but also as a *biological system*; as being a *surface* or pair of surfaces (sea/air surface and sea bottom) but also as being a *vertically* organized space, i.e., as having an irreducible *depth* (Steinberg, 2017). "Linnaean" categorizations still lurk in the taxonomies of the different "seas" and "oceans" of the world, and contrast starkly with the unitary, one-ocean conception nowadays heralded by biophysical researchers and climatologists.

That different perspectives are alive and kicking in the negotiations about the sea can be gathered from the variety of existing policies and attitudes. Steinberg (2001) distinguished between three approaches to sea control: laissez-faire, exemplified by the principle of the freedom of the sea for all; territorialization, furthering coastal states' control into adjacent marine areas; and stewardship, a form of limited control to protect maritime interests without fully embracing territorialization. These conceptualizations compete for embodying the mercantile, navigational and military vision of the ocean respectively, born of the era of European route openings and subsequent colonization. The protection of marine areas is at times synergistic with economic interests about their content, at times antagonistic. Appropriation is a common solution, mostly based on land-projected areas (territorial waters, Exclusive Economic Zones), as if drawing lines on a map was a cognition-friendly, always at hand solution. Marine protected areas (MPAs) further the territorialization of the sea in the name of environmental protection, although in general territorialization cannot efficiently cope with the extreme mobility of currents and of the sea biome (e.g., the benefits of spill-over effects on ocean ecosystems are still widely debated). Recognition of the biological primacy of the sea has given rise to other legal protection instruments, such as the individuation of Key Biodiversity Areas (Abreu et al., 2019). On a different path, the valuation of the ecological services of the ocean and of its inhabitants is reflected in different approaches, from the monetization of (parts of) the ocean (Deloitte, 2017) to the management of commons, from the *labelization* of (parts of) the ocean as *common heritage*, according to the 1982 Montego Bay Convention, to the proposal of attributing legal personhood to the ocean (David, 2019), to the scheduling of moratoria in ocean exploitation.

What are the options for reengineering ocean-related concepts, in view of the deployment of sensible long-term policies?

4.3.4 The Perils of the (Purely) Lexical Move

Faced with the recognition of conceptual inadequacy, one possibility is to change the words used in the conversation, or to invent new terms: that "we needed new language to deal with mass habitat damage and destruction, and the loss of ecosystems" (Wright, 2022), was one of the explicit motivations for the introduction of the term *ecocide* in international treatises and national laws. We can also start by existing fully lexicalized concepts, then replace the names we use in the conversation with fitter descriptors: no longer speak of ?*fisheries* but of *provinces of extraterrestrial living beings*; intervene on ontological categories: blocking the use of "fish" as a mass term–**fish portions, *rations*–so as to focus on fish as *individual living beings*; no longer speak of **oil fields* but of *underwater carbon buffers* or of *planktonic gifts*; not of **resources* but of *survival conditions*; not—as said—of **natural gas* but of *fossil gas;* no longer of ?*fishing*, in some cases, but of *mechanical extraction.* The lexical move is widely practiced and has the undeniable cognitive advantage of suggesting that an immediate action has been taken. However, in working on reconceptualizations we should be aware of the risk that a terminological change may not lead to a conceptual replacement, what Pinker (1994) aptly called "the game of the name". A change of words is not a guarantee of conceptual change; new baptisms and terminologies, if not adequately accompanied, may just drag old concepts with them.

4.3.5 The Perils of Unsupported Metaphors

Some have called for action on the metaphors and images we use to describe the marine environment. The metaphor of the **plastic continent* floating in the Pacific, **as big as Texas*, has gone viral in the last decade. However, it seems to have proved useless and counterproductive for conservation policies: it is not well supported by scientific data and is easy to disprove when it is accompanied in the media by misleading and defeasible images (plastic debris close to shore); it also instills the idea of the prevalence of a certain type of pollution (plastic packaging, bottles, bags) and hides others, invisible but nonetheless important, that we cannot afford to underestimate. (Neilson, 2018) suggests that metaphors, to be effective, should draw from a body of vivid intuitive knowledge, but that in order not to be counterproductive, they cannot forgo scientific accuracy. Among the examples of metaphors that balance vividness and accuracy Neilson proposes there is talk of the *osteoporosis of the sea* (i.e., ocean acidification makes shells brittle), of *corals as lungs* (filters), or saying that deep-sea sound pollution creates a *background fog* for communication between cetaceans, a fog in which ships' sonar pulses explode like *dynamite*. One may also present the rising sea temperature as a form of *sea fever*. Taken further, we can leverage our knowledge of human fever mechanisms and point to how a 1 °C difference in body temperature, in itself inconspicuous, can become lethal above a certain threshold. You do not want your body to increase 1 °C if your temperature is already 40 °C. It may further help not to think of beaches as either pristine paradises or open-air garbage bins as much as *combs* or *sieves*, mobilizing the image of childhood play on the beach, and emphasizing a functional role of the beach in the marine ecosystem.

4.3.6 Beyond Food

The living sea enters our homes daily in the form of *canned animal protein*. We can reconceptualize the catch; it would help us not to start, in our daily perspective on the sea, from canned tuna, the prototype of food rations, which makes us see living fish as just an initial stage in the packaging process (Fagan, 2017). One opposing narrative is to hone kinship links with the natural entities we feed on. "The salmon and the crabs and the clams are relatives. They're living relatives. They're not just resources. And so, you treat them with a symbiotic respect. They feed you because you take care of them. It's a very different way of thinking about why these areas are important", testifies environmental scientist Jamie Donatuto, who serves as the environmental health delegate for the Swinomish in Washington State, self-described as the salmon people (Morrison, 2020). Which conceptual resources can we mobilize here? We know that it is possible to create *kinship* beyond the immediate genetic link, beyond biological filiation: the explicit *adoption* of a son or a daughter, an ancient universal and conceptually unproblematic practice, and the implicit adoption when a family is recomposed, bring with them a fabric of relationships that, if they are not biological, are not just simple friendship or temporary cohabitation.

At a minimum, a quantitative indicator could be attached to marketed seafood, indicating the proportion of the catch that reaches the shelves to that of the discarded catch (FAO, 2005). More ambitiously, perhaps, we could wrap narratives around the complex biological and evolutionary history of each species, about the intriguing *cognitive abilities* of some marine species. Octopuses would not be our food, but our *teachers* (Godfrey-Smith, 2016; Thys, 2020). If evolution is given proper place in narratives, as a builder not just of species, but of complex ecosystems that are like magnificent artworks or exquisitely carved temples, mechanized-industrial **fishing* and other extractive practices would be best reconceived as forms of *looting*.

4.3.7 The Scale Issue; Salted Water and Clear Soup

As an environment, the ocean is way too big at the human physical and cognitive scale, while being counterintuitively small at the planetary and astronomical scale. As a matter of fact, the sea is in a conceptual limbo, an entity halfway between the dimensions of the mesoscopic landscape (e.g., valley, mountain, lake) and that of astronomic object. Notably, the surface of the ocean is about ten times that of the Moon. Indeed, the ocean is too big to conceptualize on many dimensions that are not only geographical, but also social, biological, technological, anthropological, economic, and of familiarity; this is only amplified by the absence of a single referent community, and by the presence of a myriad of conflicting interests.

Science-based reconceptualizations are often too technical to command action, e.g. as they involve the mathematical and computational models that underpin the notion of the ocean as a unitary dynamical system (Chidichimo et al., 2023), but some of them are both inescapable and accessible. Some current notions strongly affected by scientific knowledge witness a general dynamic of replacing commonsense concepts with scientifically informed concepts. Of course, the move is constrained by the understandability of the new concepts and their ability to displace the commonsense variety. A large reconceptualization of seawater as entities living in it has been underway in science in the last century. That the ocean is a **container* of living entities, *that it* is **just water and winds*, with fish, birds, and whales, that sea-water is **just salted water*: all this has given way to an idea of the sea as a *(quasi) living entity*. Seawater is better seen as *a clear soup* (Selosse, 2017),[4] it has been *the theater of life's origin* and today it produces half of living matter; it contains invisible planktonic forests, but some of the underwater vegetal items are larger than many land trees. In parallel, considering sea living beings as **alien living beings* is strongly contrasted by the Darwinian highlighting of *kinship relations*: diatoms are *distant cousins*. Biota from the entire tree of life are found in every liter of seawater; these form ecosystems which can be seen as invisible, microscopic floating forests. On the other hand, we, humans, are flooded by blood whose salt composition is similar to sea-water; we bear in our inner body the memory of the ocean and ocean life that has evolved in seawater for more than three billion years. Moreover, the notion of the ocean as a **monotonous, homogenous global biome*, gives way to the fact that although features in the ocean are typically more flexible, dynamical, and boundless than the ones on land, the world ocean has a specific, complex biogeography, with distinct provinces and seascapes (Richter et al., 2019).

Science also moved from **streams are purely physical phenomena* to *streams are biological conveyors*, from **sea waste is a set of objects* to *sea waste is a process* (Tani, 2020), from **icebergs are chunks of matter* to *icebergs are frozen ocean life*,[5] from **plankton is food for whales* to an articulate conception of plankton

[4] "The waters around us are . . . like a clear muddle of microbes...which account for 90% of the total marine biomass" (Selosse, 2017, p. 318).

[5] Icebergs also preserve aerosols such as iron, which can fertilize seawater when they melt.

as an invisible multitude, an extraordinary global ecosystem that generates half of planetary oxygen, sustains all marine life, and regulates climates (Sunagawa et al., 2020). Most importantly, ocean science has disposed of old, natural historical classifications of **many seas and oceans*, (e.g., the **seven seas*, the **five oceans*) and converges on the idea that there is only *one ocean*. This goes hand in hand with a move from the idea that **human intervention has mostly local consequences*, thus global consequences are overlooked (regionality), to the idea that *human intervention is holistic*.

4.3.8 Spatializing and de-Spatializing the Sea

The sea is structured as a mass-entity: in the open sea every portion of water is at least *prima facie* fungible. The conceptualization of any part of the sea is tributary to a terrestrial datum precisely because a mass of water does not stay in the same place and does not preserve its spatial cohesion over time. In the unresolved dialectic between place and matter, so typical of the sea, we cannot focus on the place without losing sight of the matter, or on the matter without ending up abandoning the place. At most, as a substitute, we will focus on a region of the seabed, or to a part on the surface above a well identified and mapped seabed; the logic of protected areas, born to protect the land, extends, not without difficulty, to marine areas. We understand relatively well surfaces,[6] but the sea surface and the sea bottom are (somewhat) mappable. Less so, for the time being, sea volumes, although the deployment of the cubic kilometer neutrino observatory structures will create physical structures that define stable 3-D spaces in the depths of the Mediterranean.

4.3.9 On the Coastline: Rethinking Boundaries

In discussions about the law, it has been long recognized that a conception according to which **the land dominates the sea* (both conceptually and politically), or **sea geography is projected from the land*, should be scrutinized, in order to set the sea free from the land, or at least that we acknowledge that sea and land are interrelated. Territorial waters and Exclusive Economic Zones are drawn starting from coastlines through the application of a set of geometric rules. Thus, these legal concepts create interesting geopolitical facts. Take France, for example. France has a relatively small land territory, and yet it turns out to be the second-largest maritime domain because of the application of the algorithm for drawing maritime boundaries.

We can rethink the land-sea interrelation from the perspective of the sea. Land elevation provides a start case in which the land is conceptually tributary of the

---

[6] However, see (Stroll, 1988).

sea, when it is measured as Height Above Mean Sea Level. The latter is based on a determination of historical sea level, which in turn depends upon sea volume—which is also changing itself because of climate oscillations—and land subsidence or uplifting (Labeyrie, 2015). Thus, a certain mutual conceptual dependency of land and sea is acknowledged in the measures for elevation above sea level.[7]

In general, a lot of negotiation focus needs to be put on boundaries. Reconceptualized boundaries can address inadequate representations of the ocean that need to change for more adaptive and sustainable legal strategies. Even accepting that there are no strict physical "boundaries" on the land, it is clear that boundaries in the sea are hard to maintain and police, as opposed to boundaries on land (although the latter also deserve scrutiny (Smith, 1995). Allegedly, the only intuitively strong marine boundary is the boundary of the sea with land (Carson, 1955), but here too the situation is complex and configures an interesting *limit* case. The notion of **coast as a boundary, an edge, a barrier* (i.e., a two-way boundary, for both marine and inland dwellers, a static notion) could be mitigated by a notion of coast as *filter, porous exchange membrane*, or of coast as *negotiated margin* (i.e., a procedural notion); similarly, the notion of **coast as the referent of a line on a map* is replaceable by an idea of *coast as integrated area* involving both land and sea. Thus, a static representation of the coast—implying various misrepresentations of coastal over-consolidation, protection, demarcation—could give way to a representation of the coast as dynamic, intrinsically changing, on human life-compatible temporal scales. Accordingly, the idea that **all coastal portions are born equal* can be replaced by a form of "conceptual biodiversity," that pays attention to marshes, lagunae, deltas, cliffs, ports. Consequences of the new notion will spread to coastal actors: the claim that **coastal real estate is the same* as any other type of real estate (same protection level, insurance, rights, as any other type of inland estate) will be modulated by specificities of coastal real estate; the claim that **posidonia is pollution* could be tempered by the aspect of *posidonia as consolidator*; the claim that **waves are attackers of the coast* will be integrated by reflections on *waves as foci of cultural identities* (e.g., by the communities of surfers), and on waves as *energy generators*.

4.3.10 Parametric Change: Time

Conceptual engineering proceeds sometimes by *parametric change* (Casati, 2011). The conceptual negotiation can introduce mechanisms of protection and defense, not (or not only) geographical, but (also) temporal. We can try and go beyond the idea of the sea as a *space to be preserved*; necessarily, this space will always end up being a *portion*, often coastal, of the sea-space. Instead, by parametric inversion,

[7] Other measures of elevation are ocean-independent, such as the distance from the center of Earth or geocentric distance, which makes the Chimborazo "taller" than Mount Everest because of the Equatorial bulge.

we can think of a different organization of *time* to be able to invest *all* the space of the sea. Regulators could propose from a periodic blank year, or even just a blank month, in which our species doesn't go to sea, for example as early as 2030, providing compensation for those who live off the sea. Empirical evidence provides some encouragement. During the 2020 lockdowns, whales have moved closer to the coast (Leicester, 2021), probably no longer disturbed by the noise of boats, and at the same time industrial fishing has declined. We know that during the two world wars of the twentieth century, the reduction in fishing activity meant that fish stocks were able to rebuild to some extent. Parametric switch would imply that instead of protecting **a limited sea space for an indefinite time*, we *protect the whole sea for a limited time*.

4.3.11 Beyond Language: Iconicity, Perspectivality, and Narrative Moves

Some reconceptualizations–as has transpired from the above–may require abandoning lexical, definitional, or parametric focus, and move to a yet different type of game. Change of perspective (the "Kepler move") and non-linguistic representations (images, maps), and in some cases both, may induce new ways to apprehend the ocean. As said above, the iconic "earthrise" picture played a pivotal role in rising environmental awareness. Non-standard cartographies such as West up (Réthoré, 2011), McArthur's South up Corrective Map of the World, the Spilhaus projection (Spilhaus, 1983; Spilhaus, 1942) highlight and possibly correct biases that may reinforce stereotypes and influence decision-making. The Spilhaus projection represents the ocean as surrounded by land and instills a much sought-after sense of unity of the ocean. Narrative modes of presentation of physical facts can intervene as well. Resized at the scale of the Earth, a chicken egg would be ten times rougher than the Earth. This means that the combined width of the atmosphere and of the oceans, at the scale of the egg, would be that of an almost imperceptible humid foil—which we may call the *biofoil*.

4.3.12 Analogy Mining: Insurance, Citizenship, Legal Personhood, and Plankton Unions

Recent reconceptualization proposals have involved the reworking of human- and social-born notions to rethink natural and maritime entities. The proposal is not simply to translate human-centered notions to the ocean and the biome, but to work flexibly in the attempt to overcome the stiffening of concepts, as happened with the nature/culture opposition (Descola, 2013).[8] For instance, nature itself can be thought

[8] However, if there is a place where the nature/culture opposition is recalcitrant to deconstruction, this is the ocean.

of as a form of *life-insurance* (Frère-Escoffier, 2020). *Citizenship* (Squarcina & Pecorelli, 2017) is another source domain for analogy mining. We understand what it is to be a citizen of a *country* (a series of entitlements and responsibilities flow from the attribution of citizenship) and we can work out *ocean citizenship* (a global citizenship with rights and duties of every human being) as an extension of the concept. The sea is here conceived of as an institutional structure rather than a biophysical or geographical entity, and sea citizenship as a process that involves individual appropriation of duties that mirror existing entitlements: as we depend on the sea, we should in return engage with it responsibly.

Likewise, reconceptualizations of the ocean as the bearer of legal personhood have been proposed (David, 2019). Can we create a conceptual framework that envisions the extension of personal rights to maritime entities and, eventually, to the ocean? Extension of rights and of personality in its early, breakthrough phases is always considered problematic, to the point of being deemed "unthinkable" (e.g., conferral of rights and personality to women, to colonized populations, to slaves, to the young and the elderly, to the poor; but also, to man-made institutions such as corporations, organizations, states, and churches). Recently, non-human entities like animals and trees, rivers and lakes have been granted rights. The implicit social contract that underpins our living-together is thus extended towards a contract which includes nature as a partner that is to be treated according to principles of reciprocity and respect (Serres & McCarren, 1992). If many generic difficulties and negative externalities plague some if not all these proposals, specific problems are related to the very nature of the ocean. For instance, the attribution of *legal personhood*, or the management of *commons*, when meant to counter the idea of the ocean as an **extractive resource*, appears to presuppose, in order to be effective, a relatively circumscribed spatial and social referent community, with easy to individuate common interests; by contrast the sheer size and otherness of the ocean, and the presence of literally thousands of (conflicting) actors—peoples, corporations, interests, organizations—renders those instruments hard to implement.

Diplomacy is another recently analogically mined domain (Morizot, 2020). Short of nominating a full-blown representative of the sea, who may sit in constitutional assemblies or have a standing in courts, the minimal attitude is to take other living beings' perspectives and, in the first person, give them a *voice*. This could count as a pragmatic rather than ontological reconceptualization: different courses of actions are made possible, and some of our actions will be constrained by these new action paths.

Analogy mining could be made richer and more complex. If a sea entity like plankton is emotionally remote, alien in its aspect, and too distant a cousin on the evolutionary tree, making it difficult for us to take its perspective and provide it with a voice, it is instead *conceptually* very close to us on another front that the scientific literature has recently highlighted. Plankton *works* for us, sequestering carbon and generating oxygen, producing oil and whole mountains that are quarried: it renders a planetary *service*. Attempts at conceptualizing nature's work have previously been made, for example by the United Nations who labeled them **ecosystem services* (IPBES 2020), a concept which is criticized for its anthropocentrism and exploitative implications (Schröter et al., 2014). Thinking about them instead as *labor*, which brings with it, at least for those who live out of their work, a rich set of images and concepts, of possible inferences and projections. Workers have *rights*, protected by *labor law*. If we fail to promote sea citizenship or giving legal personality to the ocean, we can at least establish the *Plankton Union*, creatively committed not so much to defending the survival or conservation of marine biology, as to promote the *dignity* of the work carried out tirelessly by the multitude of species and individuals that are part of the plankton. Dignity that for us is manifested in institutional recognition[9], but also, quite practically, in paid holidays, limited working hours, a fair wage, maternity and paternity leaves, weekly rest, vacations, open-ended contracts, non-discrimination, safety regulations, pensions, and protection from harassment, intimidation, and arbitrary dismissal: all things that seem (almost) taken for granted but that not even too many years ago were so *unthinkable* that today we should remind ourselves that they are fragile achievements for us too. A labor union's purpose is to negotiate working conditions and to make them explicit in the form of contracts. Plankton, too, can only work under certain conditions—if ocean acidification reaches higher levels, carbon sequestration may be compromised. The idea of a plankton union could make its claims more salient and leverage *solidarity among workers across species*, as well as our knowledge of what it means to work and in particular to work with dignity, so as to propose not only predictable measures such as seasonal rest, or reproductive leave, but also new measures such as the portability of a set of rights across geographical areas that are now regulated by the nations that have them. Plankton is the migrant par excellence, stateless, difficult to pigeonhole. If today's labor requires more and more mobility without adequate protection and heads towards turning all of us into migrants with only minimal rights, legal imagination can create instruments that defend all workers, on land as well as under the surface of the ocean; and if what applies to us can already apply to plankton, what we put in place for plankton can also protect us one day (Casati, 2021).

4.4 Blueprint for in-Depth Research: Limit Cases

We listed some issues: the complexity of dealing with or proposing an all-encompassing concept of *ocean*; restraint from endorsing a polysemantic account, that would imperil interdisciplinary conversation and the articulation between research, culture and policy making; the inherent inefficiencies of lexical change; the existence of non-lexicalized concepts. In view of those issues, we can envision

[9] Cf. article 1 of the Italian Constitution: "Italy is a democratic Republic founded on labor", rather than on inherited privileges. Labor is given a supreme dignity by the lawmaker.

a different approach, which tackles *limit cases* and boundary conditions. As in the study of mathematical functions one can usefully focus on extrema and notable points, in conceptual negotiation one can work on limit cases that put the conceptual structure under some pressure and elicit its scaffolding.

A list of promising candidates for conceptual negotiation at the limits includes the following: Icebergs, Ocean Currents (some of which have been discovered as late as of 2010, see (Fukamachi et al., 2010)), Coastlines, Winds, Waves, Clouds (Hamblyn, 2001), Hurricanes (NOAA, 2021), the planned Creation of seas in the Sahara desert, Straits (Harvard Law Review, 2020), Point Nemo, Mediterraneans (i.e., there are plenty of them, on top of the eponymous one), and Artificial Islands in the high seas.

To mention three of them:

1. Icebergs are an interesting test-bed for assessing conflicting intuitions (Remaud, 2020). Some are sea-born, some are land-born. When at sea, are those that are born as portions of land glaciers *floating portions of land* or *portions of the sea* instead? The presence of icebergs on newly opened commercial routes will increase due to global warming. It has been suggested that they can be used as freshwater resources. If tugged, they may fall under the tugging vessel's jurisdiction, no matter where they originate from. If free floating, are they owned by the countries in whose waters they float, or to the country they originated from? To what extent can we refine the notion of an iceberg to cover cases of attrition?

Icebergs give rise to two sorts of inconsistencies: first, and predictably, between how people categorize icebergs and their legal status (but also because the legal status is complex and the main point is that icebergs acquire variable legal statuses according to their origin and current location), and second, more interestingly, within cognitive dimensions themselves. By the latter, we think in particular of the conflict *resource/not resource*, and within the resource category, whether it is something to *mine* or to *catch* (actually although it is a mineral resource, we suspect people see an iceberg more like a whale than like gold); and finally icebergs could be categorized in terms of the seemingly paradoxical concept "naturefacts," as both natural and artificial things (Heersmink, 2013).

2. Point Nemo is the furthest place in the sea from any landmass, located approximately halfway between New Zealand and Cape Horn, between the Equator and the South Pole, at 48°52'S and 123°23'W, more than two thousand five hundred kilometers from the nearest island. In fact, Point Nemo is the meeting place of land and sky, at least from the perspective of international policies and agreements. Quite prosaically, the agencies of various states have agreed to shell a region centered on Point Nemo with the still more or less controllable wreckage of various orbiting objects, including the entire MIR space station, and a plethora of artificial satellites, a total of between two and three hundred pieces of space junk that have undergone splashdown. The status of these activities is conceptually interesting because they can be *both lawful and unlawful*, according to different international laws (Spini, 2020), (De Lucia & Iavicoli, 2018), and they put pressure on the notions of space, disputed between ocean space and outer space.

3. Waves suffer from an inadequate naive representation. They are considered as unitary objects by commonsense, and have an individual identity for technically competent surfers; but metaphysicians have ranked them under the category of *disturbances*, which encompasses knots as well (Karmo, 1977): dependent individuals whose identity and survival conditions are framed in terms of the presence of some shape and dynamical factors. Scientific representations of waves introduce only partially successful correctives, resulting in the idea that waves do not transport matters. Considered as **mere disturbances*, waves are disposable in discussions about coastal protection where, on the other hand, they provide strong, identity-forming anchors for coastal communities (Whyte, 2019). Considered as **threats* to the coast, waves call for defensive measures that alter ecosystems and prevent coastal regeneration.

4.5 Conclusions and Future Work

Reconceptualizing the ocean addresses and faces important challenges. If, on the one hand, the ocean is wrongly taken for granted as a boundless resource for extractive approaches (movement, use of water, of sea-bottoms, of biomass, carbon and heat capture, pollution dump), then, on the other hand, because of its physical and biological features, location and sheer scale, absence of an obvious human constituency or presence of conflicting constituencies, the ocean does not fit easily into innovative approaches in the law such as the attribution of legal personality or constitutional rights, as well as in education such as difficulty with individual engagement, remoteness and in business practices such as sustainable development goals.

Assessing whether reconceptualizations are viable (i.e., scientifically adequate, cognitively acceptable, and implementable) requires a specific investigation. Philosophy has been land-focused, although it could be argued that many epistemological problems have arisen from the practice of navigation (Casati, 2022b). It is perhaps time, after post-colonial and post-modern rethinking, to engage in a post-terrestrial philosophy. In an ideal research program, reconceptualizations should be tested against the replaced concept, in terms of their effectiveness, cost, and implementability, on the blueprint of limit cases such as the ones we enlisted.[10] Here we started from showing that existing concepts are not ineluctable, but above all that we can go beyond the criticism and deconstruction of existing concepts: marine reconceptualizations can and should be engineered. We hope this work will further the conversation and the de-invisibilization of the ocean.

[10] At the "Environment: Concepts and Norms" team of the Institut Nicod an empirical study of intuitions about icebergs is under way.

Acknowledgments EELISA OPE-2021-0182 (Casati, Flamme), CNRS MITI OOSP (Casati, De Vargas), EU Horizon 2020 grant No 862923 AtlantEco (De Vargas). Special thanks to Gilles Mirambeau.

References

Abella, M. K. I. L., Molina, M. R., Nikolić-Hughes, I., Hughes, E. W., & Ruderman, M. A. (2019). Background gamma radiation and soil activity measurements in the northern Marshall Islands. *Proceedings of the National Academy of Sciences, 116*(31), 15425–15434. https://doi.org/10.1073/pnas.1903421116

Abreu, A., Bowler, C., Claudet, J., Zinger, L., Paoli, L., Salazar, G., Sunagawa, S., & Troublé, R. (2019). Scientists warning on the interactions between ocean plankton and climate change. *Fondation Tara Océan*. https://oceans.taraexpeditions.org/wp-content/uploads/2019/12/Policy_brief_TaraOceanFoundation_Plankton.pdf

Anderson, D. L. (1969). Kepler's Somnium: The dream, or posthumous work on lunar astronomy. Trans. with a commentary by Edward Rosen. University of Wisconsin Press, 1966. *Renaissance Quarterly, 22*(1), 40–42. https://doi.org/10.2307/2858979

Benabou, R., & Tirole, J. (2011). *Laws and Norms* (No w17579; p. w17579). National Bureau of Economic Research. https://doi.org/10.3386/w17579.

Blandin, P. (2020). *Biodiversité: L'avenir du vivant [Biodiversity: The future of the living]*. Albin Michel.

Bourgeois-Gironde, S. (2020). *Etre la rivière: Comment le fleuve Whanganui est devenu une personne vivante selon la loi [Being the River: How the Whanganui River Became a Living Person According to the Law]*. PUF.

Boyle, C., Gray, J., Arblaster, J., Bettio, L., Webster, R., Morgan, R., & Torok, S. (2018). Earthrise, a photo that changed the world. *The Conversation*. http://theconversation.com/earthrise-a-photo-that-changed-the-world-109009

Bravo, G., & Moor, T. D. (2008). The commons in Europe: From past to future. *International Journal of the Commons, 2*(2), 155. https://doi.org/10.18352/ijc.98

Carson, R. (1951). *The sea around us* (Vol. 4, pp. 22–23). Oxford University Press.

Carson, R. (1955). *The edge of the sea*. Houghton Mifflin.

Casati, R. (2011). *Prima lezione di filosofia*. Laterza. English translation (2023) *Philosophy as Conceptual Negotiation*. https://hal.science/hal-03937758

Casati, R. (2013). *Contre le colonialisme numérique*. Paris: Albin Michel.

Casati, R. (2021, mars 28). Mettersi nei panni dei pesci [Putting ourselves in the shoes of the fish]. Il sole 24 Ore, 11.

Casati, R. (2022a). *Oceano. Una navigazione filosofica* [The Ocean: A Philosophical Navigation]. Einaudi.

Casati, R. (Ed.). (2022b). *The sailing mind*. Springer.

Chidichimo, M. P., Perez, R. C., Speich, S., et al. (2023). Energetic overturning flows, dynamic interocean exchanges, and ocean warming observed in the South Atlantic. *Communications Earth & Environment, 4*, 10. https://doi.org/10.1038/s43247-022-00644-x

Clayton, S. (2020). Climate anxiety: Psychological responses to climate change. *Journal of Anxiety Disorders, 74*, 102263. https://doi.org/10.1016/j.janxdis.2020.102263

Connery, C. (2006). There was no more sea: The supersession of the ocean, from the bible to cyberspace. *Journal of Historical Geography, 32*(3), 494–511. https://doi.org/10.1016/j.jhg.2005.10.005

David, V. (2019). *Towards a regional convention on the rights of the pacific ocean as a legal entity*. 'One Ocean' Symposium 2019, New York, New York. https://www.griffith.edu.au/__data/assets/pdf_file/0031/848722/Victor-David-Regional-Convention-on-Rights-of-the-Pacific-Ocean-as-a-legal-entity-24Aug19.pdf

De Lucia, V., & Iavicoli, V. (2018). From Outer Space to Ocean Depths The 'Spacecraft Cemetery' and the Protection of the Marine Environment in Areas Beyond National Jurisdiction (SSRN Scholarly Paper ID 3153458). *Social Science Research Network*.https://doi.org/10.2139/ssrn.3153458
de Vargas, C., Pollina, T., Romac, S., Bescot, N. L., Henry, N., Berger, C., Colin, S., Haëntjens, N., Carmichael, M., Guen, D. L., Decelle, J., Mahé, F., Malpot, E., Beaumont, C., Hardy, M., The Planktonauts, the P. P. team, Guiffant, D., Probert, I., Gruber, D. F. et al. (2020). Plankton Planet: 'Seatizen' oceanography to assess open ocean life at the planetary scale. BioRxiv, 2020.08.31.263442. https://doi.org/10.1101/2020.08.31.263442.
Deloitte. (2017). At what price? The economic, social and icon value of the Great Barrier Reef. *Deloitte Access Economics*. https://www2.deloitte.com/content/dam/Deloitte/au/Documents/Economics/deloitte-au-economics-great-barrier-reef-230617.pdf
Descola, P. (2013). *Beyond nature and culture*. University of Chicago Press.
Drouet, I., Andler, D., Barberousse, A., & Jebeile, J. (2021). Expert reports by large multidisciplinary groups: The case of the international panel on climate change. *Synthese, 199*(5–6), 14491–14508. https://doi.org/10.1007/s11229-021-03430-y
Dupont, S. (2017). I am the ocean—Arts and sciences to move from ocean literacy to passion for the ocean. *Journal of the Marine Biological Association of the United Kingdom, 97*(6), 1211–1213. https://doi.org/10.1017/S0025315417000376
Fagan, B. M. (2017). *Fishing: How the sea fed civilization*. Yale University Press.
Falkowski, P. (2012). Ocean science: The power of plankton. *Nature, 483*(7387), S17–S20. https://doi.org/10.1038/483S17a
FAO. (2005). *Discards in the World's Marine Fisheries—An Update*. https://www.fao.org/3/y5936e/y5936e00.htm#Contents
Fisher, R., Ury, W., & Patton, B. (1997). *Getting to yes: Negotiating an agreement without giving in* (2nd ed.). Arrow Business Books.
Frege, G., & Thiel, C. (1986). *Die Grundlagen der Arithmetik: Eine logisch mathematische Untersuchung über den Begriff der Zahl [The foundations of arithmetic: A Logico-mathematical enquiry of the concept of number]*. F. Meiner. http://archive.org/details/diegrundlagende00freggoog
Frère-Escoffier, L. (2020). Préserver la biodiversité [To Preserve Biodiversity]. *Etudes Marines*, 32–37.
Fukamachi, Y., Rintoul, S. R., Church, J. A., Aoki, S., Sokolov, S., Rosenberg, M. A., & Wakatsuchi, M. (2010). Strong export of Antarctic bottom water east of the Kerguelen plateau. *Nature Geoscience, 3*(5), 327–331. https://doi.org/10.1038/ngeo842
Godfrey-Smith, P. (2016). *Other minds: The octopus, the sea, and the deep origins of consciousness* (1st ed.). Farrar.
Grotius, H., (2012). Mare liberum.
Hamblyn, R. (2001). *The invention of clouds: How an amateur meteorologist forged the language of the skies*. Farrar.
Hartshorne, T. L. (1986). Modernism on Trial: C. Brancusi v. United States (1928). *Journal of American Studies, 20*(1), 93–104. https://www.jstor.org/stable/27554707
Harvard Law Review. (2020). *The potential-use test and the Northwest Passage*. https://harvardlawreview.org/2020/06/the-potential-use-test-and-the-northwest-passage/.
Haslanger, S. (2000). Gender and race: (what) are they? (what) do we want them to be? *Noûs, 34*(1), 31–55. https://doi.org/10.1111/0029-4624.00201
Hawthorne, D., & Minot, F. (1954). *The inexhaustible sea*. Dodd, Mead, and Company.
Hayes, P. (1985). Naive physics I: Liquids. In J. R. Hobbs & R. C. Moore (Eds.), *Formal theories of the common sense world* (Vol. 1). Ablex Publishing Company.
Heersmink, R. (2013). A taxonomy of cognitive artifacts: Function, information, and categories. *Review of Philosophy and Psychology, 4*(3), 465–481. https://doi.org/10.1007/s13164-013-0148-1

Hermitte, M.-A. (2011). La nature, sujet de droit? Annales [Nature, a subject of law? Annals]. Histoire, Sciences Sociales, *66e année*(1), 173–212. https://www.cairn.info/revue-annales-2011-1-page-173.htm

Holm, P., & Macdiarmid, A. (2017). Food from the oceans: How can more food and biomass be obtained from the oceans in a way that does not deprive future generations of their benefits? *SEAPEA Evidence Report No. 1*. Science Advice for Policy by European Academies. https://www.academia.edu/35298769/Food_from_the_oceans_how_can_more_food_and_biomass_be_obtained_from_the_oceans_in_a_way_that_does_not_deprive_future_generations_of_their_benefits

IPBES. (2020). *The intergovernmental science-policy platform on biodiversity and ecosystem services (IPBES)*. IPBES. https://www.ipbes.net/

Isaac, M. G., Koch, S., & Nefdt, R. (2022). *Conceptual engineering: A road map to practice*. Philosophy Compass. Philosophy Compass. https://doi.org/10.1111/phc3.12879

IUA. (2006). Definition of a planet in the solar system. Résolution No B5. https://www.iau.org/static/resolutions/Resolution_GA26-5-6.pdf

Jacques, G., Tréguer, P., & Mercier, H. (2020). *Oceans: Evolving concepts*. ISTE Ltd / John Wiley and Sons Inc..

Karmo, T. (1977). Disturbances. *Analysis, 37*(4), 147–148. https://doi.org/10.1093/analys/37.4.147

Krugman, P. R. (2020). *Arguing with zombies: Economics, politics, and the fight for a better future* (1st ed.). W. W. Norton & Company.

Labeyrie, L. (2015). *Submersion: Comment gérer la montée du niveau des mers [Submersion: How to Manage Rising Sea Levels]*. Odile Jacob.

Leicester, J. (2021, April 20). Whales filmed having whale of a time during lockdown. *AP NEWS*. https://apnews.com/article/understanding-the-outbreak-marseille-virus-outbreak-health-national-parks-10b208e805ae5e3379e67bb37054e90b

Lieutaud, T. (2014). « Il y a trois sortes d'hommes : Les morts, les vivants, et les gens qui vont sur la mer » (Aristote) ["There are three sorts of people: those who are dead, those who are alive, and those who are at sea" (Artistotle)]. *Annales Françaises d'Anesthésie et de Réanimation, 33*(3), 196. https://doi.org/10.1016/j.annfar.2014.02.001

Lovelock, J. E. (1990). Hands up for the Gaia hypothesis. *Nature, 344*(6262), 100–102. https://doi.org/10.1038/344100a0

Maccanico, Y. (2019). Italy's redefinition of sea rescue as a crime draws on EU policy for inspiration. *Statewatch*, 13. https://www.statewatch.org/media/documents/analyses/no-341-italy-salvini-boats-directive.pdf

Millock, K., & Nauges, C. (2010). Household adoption of water-efficient equipment: The role of socio-economic factors, environmental attitudes and policy. *Environmental and Resource Economics, 46*(4), 539–565. https://doi.org/10.1007/s10640-010-9360-y

Morizot, B. (2020). Manières d'être vivant: Enquêtes sur la vie à travers nous[ways of being alive: Investigating life through us]. *Actes Sud*.

Morrison, J. (2020). An ancient people with a modern climate plan. *Washington Post*. https://www.washingtonpost.com/climate-solutions/2020/11/24/native-americans-climate-change-swinomish/

Neilson, A. (2018). Considering the importance of metaphors for marine conservation. *Marine Policy, 97*, 239–243. https://doi.org/10.1016/j.marpol.2018.03.019

Nichols, W. J. (2015). *Blue mind The Surprising Science that Shows How Being Near, In, On, or Under Water Can Make You Happier, Healthier, More Connected and Better at What You Do*. Back Bay Books.

Nisa, C. F., Bélanger, J. J., Schumpe, B. M., & Faller, D. G. (2019). Meta-analysis of controlled trials testing Behavioural interventions to promote household action on climate change. *Nature Communications, 10*(1), 4545. https://doi.org/10.1038/s41467-019-12457-2

NOAA. (2021). *Why do We Name Tropical Storms and Hurricanes?*https://oceanservice.noaa.gov/facts/storm-names.html

Nordquist, M. H., Moore, J. N., and Long, R. (Eds.). (2018). Legal order in the world's oceans: UN convention on the law of the sea. Center for Oceans law and Policy, 21. *Brill*. https://doi.org/10.1163/9789004352544.

Pinker, S. (1994). The game of the name. *New York Times*. https://www.nytimes.com/1994/04/05/opinion/the-game-of-the-name.html

Remaud, O., 2020, *Penser comme un iceberg* [Thinking Like an Iceberg]. Actes Sud.

Réthoré, S. (2011). Méditerranée sans frontières [Map].

Richter, D. J., Watteaux, R., Vannier, T., Leconte, J., Frémont, P., Reygondeau, G., Maillet, N., Henry, N., Benoit, G., Silva, O. D., Delmont, T. O., Fernàndez-Guerra, A., Suweis, S., Narci, R., Berney, C., Eveillard, D., Gavory, F., Guidi, L., Labadie, K., et al. (2019). Genomic evidence for global ocean plankton biogeography shaped by large-scale current systems [Preprint]. *Ecology*. https://doi.org/10.1101/867739

Rock, J., Sima, E., & Knapen, M. (2020). What is the ocean: A sea-change in our perceptions and values? *Aquatic Conservation: Marine and Freshwater Ecosystems, 30*(3), 532–539. https://doi.org/10.1002/aqc.3257

Salmond, J. W. (1893). *The first principles of jurisprudence*.

Scholtz, G. (2016). *Philosophie des Meeres* (1st ed.), [Philosophy of the oceans]. Mare.

Schröter, M., van der Zanden, E. H., van Oudenhoven, A. P. E., Remme, R. P., Serna-Chavez, H. M., de Groot, R. S., & Opdam, P. (2014). Ecosystem services as a contested concept: A synthesis of critique and counter-arguments: Ecosystem services as a contested concept. *Conservation Letters, 7*(6), 514–523. https://doi.org/10.1111/conl.12091

Selosse, M.-A. (2017). *Jamais seul: Ces microbes qui construisent les plantes, les animaux et les civilisations* [Never Alone: The Microbes that Build Plants, Animals and Civilizations]. Actes Sud.

Serres, M., & McCarren, F. (1992). The natural contract. *Critical Inquiry, 19*(1), 1–21. https://www.jstor.org/stable/1343752

Smith, B. (1995). On drawing lines on a map. In A. U. Frank & W. Kuhn (Eds.), *Spatial information theory a theoretical basis for GIS* (Vol. 988, pp. 475–484). Springer. https://doi.org/10.1007/3-540-60392-1_31

Speich, S. (2019). The ocean, a heat reservoir. *Ocean Climate,* 14.

Spilhaus, A. (1983). World Ocean maps: The proper places to interrupt. *Proceedings of the American Philosophical Society, 127*(1), 50–60.

Spilhaus, A. F. (1942). Maps of the whole World Ocean. *Geographical Review, 32*(3), 431. https://doi.org/10.2307/210385

Spini, L. (2020). A sustainable outer space for a future on earth. Outer space, space debris and technologies. *Agathòn, 8*(15), 160–167. https://doi.org/10.19229/2464-9309/8152020

Squarcina, E., and Pecorelli, V. (2017). Ocean citizenship. The time to adopt a useful concept for environmental teaching and citizenship education is now. *J-READING Journal of Reasearch and Didactics in Geography,* 2. http://j-reading.org/index.php/geography/article/view/168

Steinberg, P. E. (2001). *The social construction of the ocean*. Cambridge University Press.

Steinberg, P. E. (2017). Oceans and seas: Human geography. In D. Richardson, N. Castree, M. F. Goodchild, A. Kobayashi, W. Liu, & R. A. Marston (Eds.), *International encyclopedia of geography: People, the earth, environment and technology* (pp. 1–6). John Wiley and Sons, Ltd.. https://doi.org/10.1002/9781118786352.wbieg0759

Stone, C. D. (2010). *Should trees have standing? Law, morality, and the environment* (3rd ed.). Oxford University Press.

Stroll, A. (1988). *Surfaces*. University of Minnesota Press.

Sugimoto, A., Sugino, H., & Hori, J. (2022). How bountiful is the ocean? Participatory valuation of human–nature relationships in Yaeyama Islands, Okinawa, Japan. *Sustainability Science, 17*, 878–898. https://doi.org/10.1007/s11625-021-00994-z

Sunagawa, S., Tara Oceans Coordinators, Acinas, S. G., Bork, P., Bowler, C., Eveillard, D., Gorsky, G., Guidi, L., Iudicone, D., Karsenti, E., Lombard, F., Ogata, H., Pesant, S., Sullivan, M. B., Wincker, P., & de Vargas, C. (2020). Tara oceans: Towards global ocean ecosystems biology. *Nature Reviews Microbiology, 18*(8), 428–445. https://doi.org/10.1038/s41579-020-0364-5

Tani, I. (2020). Considerazioni in tema di « end of waste » tra carenti norme internazionali e attribuzione al consumatore di responsabilità altrui [Considerations on "End of Waste" Between Deficient International Standards and Attribution of Responsibility to the Consumer]. *Rivista giuridica dell'ambiente, 35*(2), 243–277.

Tasse, J., & Alex, B. (2020). Le changement climatique, facteur d'océanisation des enjeux géopolitiques? [Climate change, a factor in the oceanization of geopolitical issues?] *Etudes Marines*, 102–109.

Thys, T. M. (2020). Exploring eight-armed intelligence through film. *Animal Sentience, 4*(26). https://doi.org/10.51291/2377-7478.1658

UNESCO. (2013). *One Planet, One Ocean*. UNESCO. https://en.unesco.org/themes/one-planet-one-ocean

Vosniadou, S. (Ed.). (2008). *International handbook of research on conceptual change* (1st ed.). Routledge. https://doi.org/10.4324/9780203874813

Whyte, D. (2019). Belonging in the ocean: Surfing, ocean power, and saltwater citizenship in Ireland. Anthropological Notebooks, *25*(2), Article 2. http://notebooks.drustvo-antropologov.si/Notebooks/article/view/22

Wright, D. R. (2022). *Future Sea: How to rescue and protect the world's oceans*. University of Chicago Press.

Chapter 5
Conceptual Engineering, Semantic Tolerance and Flexibility

Genoveva Martí and Lorena Ramírez-Ludeña

5.1 Introduction

There is a real question as regards what conceptual engineering consists in, what it means to engineer concepts, and what the aims, motivation and targets of conceptual engineering are.[1] In this sense, conceptual engineering is remindful of cases in which there are some clear examples of a phenomenon (and some clear non-examples), but the final definite characterization of the phenomenon is still under dispute. In this paper we hope to provide more tools for reflection by examining some cases and raising questions as to whether they can be coherently understood as cases of conceptual engineering and if so why. We discuss two semantic phenomena that are particularly important when they affect the interpretation of the law. We examine their relation to conceptual engineering, and we discuss some social and moral aspects that alert to some of the perils of prolific engineering.

[1] See Isaac et al. (2022) for analysis and discussion of the different kinds of projects that fall under the conceptual engineering umbrella.

G. Martí (✉)
ICREA (Institució Catalana de Recerca i Estudis Avançats), Barcelona, Spain

Universitat de Barcelona, Barcelona, Spain
e-mail: gmarti@ub.edu

L. Ramírez-Ludeña
Universitat Pompeu Fabra, Barcelona, Spain

M. G. Isaac et al. (eds.), *New Perspectives on Conceptual Engineering - Volume 3*,
Synthese Library 515, https://doi.org/10.1007/978-3-031-98535-5_5

5.2 Some Background. The Distinctions

Mark Wilson (1982) and Henry Jackman (1999) have argued that there are cases in which decisions as to whether a newly encountered object is to be included in the domain of application of a term can result in the exclusion of the instance from the domain, or in its inclusion, in ways that are equally consistent with past usage. Their argument is based on a thought experiment: the members of an isolated community that live on a remote island, the Druids, use the term ‘ave’ to refer to the class of birds and the class of things that fly, which have been one and the same class on the island since the distant past. When they see planes in the sky for the first time, they classify them, without a second thought, as aves, and when they later see one landing, they modify their beliefs and conclude that not all aves are living beings. And so, the practice of applying ‘ave’ to planes catches on in the community. Had they initially seen planes on the ground, noticing that they were not living objects but not realizing that they could fly, they might have classified them, also without a second thought, as non-aves. And had they afterwards seen them taking off, they would have modified their beliefs accordingly to conclude that not all things that fly are aves. In neither of the two cases, the members of the community would see themselves as having changed what they meant by ‘ave’; they would understand that their current use is continuous with their previous use.

The traditional reaction to this story postulates that the community in one or both stories must have changed the meaning of ‘ave’. But the explanation for what such a change in meaning would consist in is less than obvious, since this appears to be a case in which both the application of the term ‘ave’ to planes, and its withdrawal, seem to cohere with the meaning of the term as it had been used by the community. Thus, the moral of the thought experiment is that the meaning and use of the word ‘ave’ in that community, prior to the encounter with planes, does not determine whether airplanes should or should not be under the extension of ‘ave’: completely accidental facts about the way in which airplanes are first encountered have quite a decisive influence on the community’s application of the term.

In past work (Martí and Ramírez-Ludeña 2021) we have characterized this phenomenon as *semantic flexibility*: a term is semantically flexible just in case at a given point in time, when the issue arises as to whether to classify a newly encountered object, a sample or a phenomenon under the term’s extension, there may be more than one course of action continuous with previous usage or, as Jackman puts it, different *alternative equilibria*.[2]

Semantic flexibility has been acknowledged by different authors. For instance, according to Joseph LaPorte, pre-scientific uses of terms are open-textured and, as science progresses, it is not determined in all cases whether a term should apply to a given sample or exemplar: “Empirical exploration uncovers many plausible

[2] For a discussion of our assessment of the further conclusions Wilson and Jackman draw from the Druids’ story, especially Jackman’s endorsement of temporal externalism, see Martí and Ramírez-Ludeña (2021), especially Section 2.

candidates" (LaPorte 2004, p. 3). Other terms may simply be inflexible: only one alternative is compatible with the meaning of the term. For instance, Sabbarton-Leary (2010) points out that the meaning of 'tungsten' is given by the description 'the element with atomic number 74' and that past, present and future uses of 'tungsten' are correctly applied only if the application accords with that description.

An interesting question is whether the changes in the history of the use of a flexible term are cases of conceptual engineering. We surmise that conceptual engineering is an intentional manipulation of the meaning of a term, or of the concept associated with it, a manipulation that may result in changes in the domain of application.[3]

The phenomenon of flexibility should be distinguished from what we (Martí and Ramírez-Ludeña 2021) have termed *semantic tolerance*, the acceptability of the cohabitation, in a speaker's community, of different uses of a term that divide roughly the same domain of application in different ways. An obvious example is the term 'fruit' that has a botanic use (that applies to tomatoes) and a culinary use (that does not).[4] The different tolerated uses may be appropriate in different contexts, and a question to raise as regards conceptual engineering is whether the choice between one or another usage, or even the creation of an entirely new usage in a given context, is a form of engineering.

The dividing lines between tolerance and flexibility are not always entirely clear and distinct. But it is important to stress that the two phenomena are different. When 'fruit' is not applied to tomatoes in the culinary use of 'fruit' there is no expansion or contraction of the domain of application for future uses of 'fruit', as it occurs in the case of 'ave'. The culinary and the botanic uses of 'fruit' are two different uses both alive in the community. When the Druids apply 'ave' to a plane they are not creating a new use that coexists with the past use of 'ave', unlike what is the case with the two uses of 'fruit'. Flexibility and tolerance are two real, and different, semantic phenomena.[5] Our question here is whether they are connected with conceptual engineering and if so in what way. In order to explore that question, it will be interesting to put the focus on legal interpretation, a domain in which questions as regards the proper application of expressions often arise.

[3] Of course, these characteristics are not meant to be a set of necessary and sufficient conditions. Engineering does not necessarily result in changes in the domain of application. Some engineering may result in a different conceptualization of the same domain. Here, our discussion will focus on cases in which what is at stake is the possibility of changes in the domain of application while, arguably, preserving the continuity of usage.

[4] In Martí and Ramírez-Ludeña (2021) we have contrasted tolerance with *strictness*, the unacceptability of the cohabitation of different uses.

[5] Tolerance and flexibility, as we conceive them, are semantic properties. They are possessed or not possessed by terms independently of time. However, flexibility and inflexibility can be observed diachronically as the possibility of incorporating a newly encountered instance arises. Tolerance and strictness, on the other hand, can be observed at a given slice of time, if there are already present uses cohabitating in the community, and also diachronically, when the question of accepting a new usage arises.

Judges are often called to interpret or re-interpret the words of the law. And some of these interpretations exhibit features that make them characterizable as instances of flexibility or tolerance (Martí & Ramírez-Ludeña, 2020). Moreover, when judges make decisions, they interpret the law in ways that sometimes appear to extend or subtly alter the criteria and even the domain of application of terms, and their interpretations have legal and social consequences. But are these interpretations or re-interpretations, arguably, also instances of conceptual engineering? Our purpose here is to examine this question, focusing on specific examples. More than trying to arrive at a yes-or-no answer in each case, we hope that our discussion will raise questions that may provide tools to refine the aim and the scope of conceptual engineering.

5.3 Tolerance. The Appeal to Different Uses of Words in the Law

As we have mentioned before, the term 'fruit' provides a paradigmatic example of tolerance. There are two uses of the term 'fruit' in the community that divide the domain of application in slightly different ways. In a purely scientific, botanic, usage, the word 'fruit' applies to tomatoes. For culinary purposes, tomatoes are not classified as fruits.[6]

The existence of these two uses became legally relevant in the, by now well known, 1893 case considered by the US Supreme Court (149 US 304) in which merchants argued that tomatoes should be exempt of duty, as other fruits. Although the Court acknowledged the legitimacy of the two uses,[7] it decided that the common, culinary use was the relevant one in that case, and found in favor of tax collectors. In the words of Justice Gray (cited in Phillips, 2014, p. 381): "in the common language of the people, . . . all these are vegetables . . . usually served at dinner in . . . the principal part of the repast, and not, like fruits generally, as dessert."[8]

[6] In our view the different usages are well established and conventionalized in the community, so they are semantic, but the line between semantics and pragmatics is at times rather fine (Martí & Ramírez-Ludeña (2016, 2020 and 2021) for a discussion of these issues). Moreover, are those different uses cases of ambiguity or polysemy? There are some subtle issues as regards the characterization of polysemy that are well beyond the scope of this paper. We are inclined to regard this case as ambiguity because the two uses of 'fruit' are lexicalized differently in other languages (for instance, in Greek –we thank Yannis Stefanou for bringing this to our attention). We have decided to use the term 'uses' in order to remain neutral on all these issues.

[7] Sometimes, the plausibility of two different interpretations associated with different uses of the same term is signaled by judges by dividing the costs of the proceedings.

[8] We apply the notion of semantic tolerance to terms. Our notion is arguably connected to the notion of *dual character*. In broad strokes, dual character concepts apply to something in some sense but not in another sense. The notion of dual character is often used to distinguish a descriptive and a normative character of concepts. See (Reuter, 2018) for a discussion of the research on dual character concepts. Here we do not engage in the discussion of what dual character consists in,

Given that the court acknowledged the legitimacy of both uses of 'fruit', things might well have gone differently. And, in fact, in some cases it is not the common use that is deemed relevant in a legal dispute, but rather a scientific or technical use. In spite of the common and extended use of terms such as 'soy milk', 'almond milk' or 'oatmeal milk', in June 2017 the European Court of Justice found, against the German company TofuTown, that plant-based products cannot be advertised or marketed "with designations such as 'milk', 'cream', 'butter', 'cheese' or 'yoghurt'." Such terms, according to EU law, are to be reserved for animal products.[9]

And yet, some cases illustrate how courts may rely at different times on different uses of terms present in the speakers' community. In the 2007, 2008 and 2009 Pringles case[10] the courts had to determine whether Pringles contained a sufficient amount of potato to be deemed potato crisps. In order to avoid paying VAT, the manufacturer, Procter & Gamble UK, had argued that their product were not potato crisps since they contained only 42% of potato. The courts first found in favor of the company, a decision that was later reversed.

The lower court that initially found in favor of Procter & Gamble UK relied on a technical use of 'made of' sanctioned by culinary experts, according to which a product that did not contain more than 50% of potato could not qualify as made of potato. But:

although we believe it is not likely that the difference between the culinary use of 'fruit' and its botanic use can be made in terms of the descriptive/normative axis. For some of the dual character concepts discussed in the literature, the normative aspect is clear, since considerations such as 'what a real philosopher/scientist *should do*/*should be* like' arise very naturally. But it is less clear that such considerations arise as regards 'fruit'; and if they do, it is not obvious that culinary uses of 'fruit' are any less normative than botanic uses. If the distinctions between the different forms of application of concepts are not made simply in terms of the descriptive vs. the normative, perhaps we might be able to open the door to acknowledge the existence of multiple, not just dual, character concepts. We are grateful to Nicolò D'Agruma and Jakub Rudnicki for a helpful discussion of these issues.

[9] See Press release 63/17 https://curia.europa.eu/jcms/upload/docs/application/pdf/2017-06/cp170063en.pdf. For the full text of the judgement of the Court see https://curia.europa.eu/juris/document/document.jsf?text=&docid=191704&pageIndex=0&doclang=EN&mode=lst&dir=&occ=first&part=1&cid=3868290 (last accessed December 20, 2022). The measure has been implemented slowly, perhaps due to the fact that most companies had a surplus of containers with the word 'milk' already printed, but the term is slowly disappearing from markets in the EU. Tobia et al. (2020) suggest that in legal disputes judges tend to favor the common usage. This case shows that this is not always the case. The use of 'milk' for plant-based products was very well established in the marketplace, and it continues to be so in common parlance. We thank Nicolò D'Agruma for bringing this case to our attention.

[10] Proctor & Gamble (UK) v Commissioners for HMRC (2007) (Tax Tribunal), Commissioners for HM RC v Proctor & Gamble (2008) (High court) and Commissioners for HMRC v Proctor & Gamble (2009) (Court of Appeal). See http://news.bbc.co.uk/2/hi/business/7490346.stm for a news release on the original decision of the lower court and https://loweringthebar.net/2009/05/pringles-qualify-as-potato-chips-court-rules.html for the reversal (last accessed December 20, 2022).

> [t]he Court of Appeal [. . .] found that the plain meaning of the term "made from" did not require the product to be more than half what it was allegedly made from. This is not a question for scientists or "culinary pedants," the court said: "[M]ost children, if asked whether jellies with raspberries in them were 'made from' jelly, would have the good sense to say 'Yes,' despite the raspberries."[11]

These cases illustrate how the uses already existing in the community are imported in the legal domain. The decisions of the courts are not meant to change the different uses of terms by contributing to eliminating or altering them. Are these cases of conceptual engineering? Certainly, the judges intentionally choose one or another use of an expression as the relevant one in adjudicating the case, but nothing that could qualify as an alteration of meaning or of concepts appears to occur. These are cases in which the uses already present in the community at large are regarded within the legal domain as legitimate, and possibly relevant for different legal decisions; they do not appear to be instances of conceptual engineering.

In those cases, the decisions of the judges do not result either in an alteration in the domain of application of the terms in the Law, as compared with how those terms are applied by experts or by people in the street, so they do not alter the linguistic practices of the community. As regards how the particular case is resolved and especially if the decision constitutes a binding precedent, the judge's decision has an impact on what the correct use is in the legal domain, and precisely because of this, it may end up having an impact on the use by the community at large.[12]

From the point of view of tolerance, cases in which judges were, arguably, acting as conceptual engineers would have to be cases in which new uses, not previously existing in the community, were created within the legal domain. For instance, had judges decided in the Pringles case that anything that tastes like a potato will be taken to be made of potato, they would be creating a new use for 'potato' not already present in the community. Are there any such cases? In principle it seems unlikely that in interpreting the law judges will, out of the blue, create a new use for an expression.[13] But it is not entirely farfetched to conceive of a scenario in which, during a dispute as regards some genetically engineered foodstuff, the courts could impose a new commercial or culinary use for a word that previously had had only one use in the community.

[11] https://loweringthebar.net/2009/05/pringles-qualify-as-potato-chips-court-rules.html (last accessed December 20, 2022). For a description and discussion of the case see Hutton (2014), pp. 73–79.

[12] For a discussion of the role of interpretive precedents in civil law systems, see Ramírez-Ludeña (2023).

[13] This does not exclude the possibility of the existence of legal gaps, situations in which judges decide *ex novo* how to resolve cases; the emphasis here is on the fact that when interpretive problems arise, the issue can be seen as a choice between pre-existing uses.

5.4 Flexibility. Extending Domains in the Law

We say that a term exhibits semantic flexibility whenever there is more than one alternative path, consistent with past usage, in which the application of the term (to a particular representative of a group) is accepted or withdrawn.

The fictional 'ave' case is a clear instance of this phenomenon, for both courses of action upon encountering planes are equally compatible with past usage. Nothing in the semantics determines the correctness of one over another choice; the decision is determined by factors that have to do with the external conditions in which the first appliers encounter the objects in question.

The 'ave' case is a story in which the alternative choices in the different situations happen naturally: no intentional process to modify the concept of *ave* is at stake in the story, although in each case, the decisions result in different domains of application for the term 'ave'. It seems unlikely that this case could be taken as an illustration of conceptual engineering.

However, other cases are quite different. As we mentioned before, Joseph LaPorte argues that, when it comes to pre-scientific uses of terms such as 'water', "it is not clear whether a new substance with the familiar superficial features but a different microstructure, or a new substance with unfamiliar superficial features but the same microstructure, would belong in the extension" (LaPorte, 2004, p. 103). Discoveries, such as the discovery of a certain molecular structure of certain samples, put the scientist, or the community in general, in the position of deciding whether the new information will be considered part of the similarity that is relevant to classify something as being the same substance or not. This is indeed an intentional decision, that according to LaPorte has been made, at least in the case of the term 'water', by scientists. And the decision to follow one alternative course of action rather than another, determining the future domain of application in a given way, is the result of an intentional process that is, arguably, we surmise, a form of conceptual engineering.

An illustration of the impact of semantic flexibility in the Law is the debate around the application of 'marriage' to same sex couples. In the decision to accept same sex marriage, we could say that the courts have treated the term 'marriage' as semantically flexible. Traditionally marriage had been conceived as a union of man and woman. At the junction when the term 'marriage' could be extended to same sex unions or remain applicable only to unions of woman and man, some legal systems decided to extend it. In some legal systems, both courses of action, would have been, arguably, compatible with prior use. For instance, the governing party of Spain modified the Civil Code to allow same sex marriage in 2005. The opposing conservative party challenged the constitutionality of the new norm and then the Spanish Constitutional Court explicitly acknowledged that either course of action was compatible with the Spanish Constitution and with what the Spanish Constitution says about marriage.[14]

If we put the issue in Jackman's terms, the Spanish Constitutional Court acknowledged that two different equilibria were consistent with the use of the term 'marriage', one of them accepting the application of 'marriage' to same sex unions, and the other one continuing to restrict it to unions of woman and man. Is the extension of the domain of 'marriage' to same-sex couples an instance of conceptual engineering?

Observe that the 'marriage' case is quite different from the 'ave' case. As we have argued above, in the 'ave' case, the Druids do not face a choice. Each of the alternative histories presents a naturally continuous application of the term 'ave'. There is no intentional choice of one alternative over the other one, which is why we have concluded that the 'ave' case, a case that paradigmatically illustrates the phenomenon of flexibility, is not an instance of engineering. In contrast, in the 'marriage' case we are faced with two alternatives, each of which, at least according to the Spanish Constitutional Court, is equally compatible with the prior usage of 'marriage' in the Law, and in society in general. All the Spanish Constitutional Court does is recognize the semantic flexibility of the term 'marriage' and the existence of two alternative equilibria. The Court leaves to the legislature the decision as to which alternative to take, a decision that is influenced mostly by social factors that are independent of semantic considerations.[15]

On the one hand, it would seem that this is not an instance of conceptual engineering, for the alternatives are seen as continuous with past usage, so there is really no manipulation of the meaning of 'marriage' or the concept *marriage*. On the other hand, the decision to take one or another course of action is clearly intentional, and the result of the alternative taken by the legislature, subsequently sanctioned as constitutional by the Spanish Constitutional Court and applied by other courts, alters the domain of application of the legal term 'marriage'. So there seems to be a manipulation that is arguably a form of engineering; a form of *collaborative* engineering, involving the legislature and the Constitutional Court. It is nevertheless difficult to say whether it is *conceptual* or not.[16]

[14] In the Spanish legal system, the Constitutional Court controls the constitutionality of norms always *ex post*. See Ramírez-Ludeña (2018) for a discussion of the same sex marriage case in Spain.

[15] Same sex marriage has been legal in Spain since 2005. Of course, since the two alternatives were semantically legitimate, things could have gone in the other direction. Because the Court acknowledges the compatibility with the Spanish Constitution of both courses of action, the legislature might, even now, change the norm that endorses same sex marriage. To put it in Jackman's terms, rejecting same sex marriage continues to be an alternative equilibrium.

[16] Observe moreover that semantic flexibility, unlike semantic tolerance, is not connected to dual character (see footnote 8). In cases of semantic flexibility, when a decision is made to apply or withdraw the application of a term to an instance, the situation would not be adequately portrayed as 'in a sense, yes, and in another sense, no'.

Here we have been treating the debate over the extension, or non-extension, of the application of the term 'marriage' to same sex couples as illustrating the phenomenon of semantic flexibility. But some of the authors involved in the legal debate have not conceived of the issue in the same terms and have taken a position that we can characterize as semantically inflexible, denying the existence of alternatives equally compatible with past usage. For instance, Brian Epstein and Adèle Mercier, both supporters of same sex marriage, have argued from positions that assume the inflexibility of the term 'marriage'.

Brian Epstein accepts that 'marriage', as used traditionally, strictly involves the union of man and woman. He sees the acceptance of same sex marriage as a declaration that the old traditional institution of marriage is defunct, and he welcomes the advent of a new institution. It so happens that we have decided to use the old name 'marriage', for a new concept and a new meaning.

> At its core—in belief, in practice, in function and in law—the old institution of marriage applied exclusively to opposite-sex couples. Which is why the old institution is now being replaced with a better one [. . .] we are making a social decision to apply an old word to a new institution. (Epstein, 2015)

It would seem that since, as according to Epstein, there are really two words, two concepts and two meanings, there is no issue here as regards engineering any concept. A newly minted concept or meaning is not the result of conceptually engineering an entirely different and unconnected previously used concept or meaning. However, if the concepts are connected, in that the concept of marriage that applied to the defunct institution and the concept of marriage that applies to the newly created one have some essential features in common, then Epstein's proposal could be characterized as an instance of conceptual engineering, as some form of continuity, we think, is expected in cases of conceptual engineering.[17]

Adèle Mercier, on the other hand, acted as expert witness in Canadian Courts, in support of same sex marriage. In (Mercier, 2007) she considers a semantic argument according to which same sex unions would always have been under the umbrella of the concept of marriage, for the characteristic features that define a marriage (commitment to a life in common, respect, trust, etc.) are present in heterosexual as well as same sex unions. Thus, it was an error not to sanction legally same sex unions as marriages. Mercier's position entails that the concept of marriage has not changed and, therefore, that we do not have here an illustration of conceptual engineering. But if conceptual engineering is seen not only as a question of altering concepts,

[17] A referee suggests that our position here is puzzling, since conceptual replacement is among the most paradigmatic strategies promoted by conceptual engineering, as Scharp (2013) argues. As we explain in the text, we suppose that some connection between the old and the new concept is presumed in the case of replacement. If there is such a continuity, indeed, as we argue, the case of marriage, as interpreted by Epstein, could be seen as an instance of conceptual engineering. We think, though, that this discussion illustrates two forms of conceptual engineering: it is one thing to repair and improve and old bridge, and it is a different one to let it crumble while constructing a different one next to it.

but also as a question of changing our understanding of those concepts, then it is arguable that Mercier's position does illustrate a case of conceptual engineering.

Mercier's position illustrates also the recognition that we may be wrong in our understanding of the concepts that we deploy and in our use of the words that express them. This is a position that has been defended by John Perry (2011). Perry has argued that the concept and the meaning of 'cruel' is now the same it was in the eighteenth century. The death penalty, Perry argues, was as cruel then as it is now, even though our *conception* of what was cruel may have changed. Hence, according to Perry, and also according to Mercier, we may well be wrong about our own concepts.[18]

Finally, observe that both Epstein and Mercier take semantically inflexible positions: according to Epstein, on past uses of the word 'marriage' same sex unions never fell, and were never going to fall, under the umbrella of 'marriage' and a change of meaning, concept and institution occurred when same sex marriage was legalized. According to Mercier, same sex unions were always really under the umbrella of marriage, and not recognizing them was the product of an error.

5.5 A Hard Case. The Use of 'Violence'

The 1995 Spanish Penal Code distinguishes *violence* from *intimidation* and from the *use of force*. Traditionally, violence has been understood by the legal system as a physical attack or assault, and intimidation as a psychological one, both against persons; use of force is directed to things ('fuerza en las cosas'). The three concepts are used to establish different sentencing levels in each of the crimes in which they figure as concepts. They are also used to typify different kinds of crimes. For example, the distinction is present in the articles on sexual assault (article # 178), that distinguishes violence from intimidation, and on robbery (article # 237), that distinguishes the three concepts.

According to article #178,

> Any act with sexual content performed with *violence*, *intimidation* or abuse of a situation of superiority or of vulnerability of the victim will be considered sexual aggression.[19]

As regards article #237, a robbery is defined as a crime in which someone intentionally

> appropriates alien objects by using *force on things* to access or leave the place where those objects are located, or uses *violence* or *intimidation* on people, be it when the crime is

[18] See also Sawyer (2020) for arguments to the conclusion that the target of conceptual engineering should be the revision of conceptions, not concepts.

[19] The original in Spanish: "[Se] consideran en todo caso agresión sexual los actos de contenido sexual que se realicen empleando violencia, intimidación o abuso de una situación de superioridad o de vulnerabilidad de la víctima." This article (passed in April 2023) modifies prior versions that also distinguished violence from intimidation.

> committed, to protect themselves when fleeing, or on those that try to help the victim or pursue them.[20]

Interestingly, the article on coercion (article # 172) uses only the term 'violence'. The article is applied to:

> Someone who, without being legitimately authorized, prevents another person *with violence* from doing something not prohibited by law, or compels them to do something they do not want to do . . .[21]

So, in principle, according to this article, if there is, for instance, only intimidation, without physical attack or assault, we could not say that there has been a crime of coercion. Many acts that we would naturally qualify as coercion involve the use of psychological intimidation, or involve the use of force on things to compel people to do something or prevent them from doing it. However, the law, in principle, could not classify them as coercion.

An obvious solution to this problem would have been to change the article on coercion adding 'intimidation' and 'use of force (on things)'. But changing the Penal Code is an extremely complex and long process that the legislature does not undertake easily. In particular, in Spain, the procedures required to change laws that affect fundamental rights, such as the articles of the Penal Code, are much more demanding than other changes to ordinary laws. It often occurs that the law is not altered and then, unavoidably, judges interpret the letter of the law in order to adequately deal with the new cases and the new realities.

So, in judicial decisions, it is possible to observe the increasing inclusion, under the umbrella of the article on coercion, of situations that are not instances of physical violence, qualifying them, de facto, as instances of violence.

The interpretation by the Spanish Supreme Court appeals to the *spiritualization* of violence, and both the legal doctrine and some judicial decisions often mention the term.[22] This is clearly a change between a prior conception of violence that was non-spiritual, towards a new one that explicitly spiritualizes it since it contemplates more than just physical violence. Ironically, when the Spanish Penal Code was approved in 1995 and also in its more recent modifications, the article on coercion, was not altered, in spite of the fact that the inconsistency with other articles that explicitly distinguish psychological intimidation, force on things and violence had been noted.

---

[20] The original in Spanish: "Son reos del delito de robo los que, con ánimo de lucro, se apoderaren de las cosas muebles ajenas empleando fuerza en las cosas para acceder o abandonar el lugar donde éstas se encuentran o violencia o intimidación en las personas, sea al cometer el delito, para proteger la huida, o sobre los que acudiesen en auxilio de la víctima o que le persiguieren."

[21] The original in Spanish: "El que, sin estar legítimamente autorizado, impidiere a otro con violencia hacer lo que la ley no prohíbe, o le compeliere a efectuar lo que no quiere . . ." Italics have been added to the translations in the main text. This article was originally in the 1944 (#496) Penal Code and in the 1973 (#496) Penal Code, and it has survived till present day without alteration.

[22] See Ragués i Vallés (2003) for an exhaustive list of items of legal doctrine and decisions by different courts that mention explicitly the spiritualization of violence.

The spiritualization of the notion of violence in recent years, is crucial in the application of the article on coercion. For instance, a 1989 sentence of the Spanish Supreme Court (STS 23 November 1989, reporting judge Delgado García) considered coercion the feinting to run someone over with a car. This type of action is typically considered a form of intimidation; but the Supreme Court, by considering it a coercion had to consider that action, de facto, a form of violence, thus subsuming intimidation under violence.

In 1981 the Mayor of El Campillo (Huelva) ordered to cut the supply of water to neighbors that had not paid a specific tax. The neighbors sued the Mayor and the Supreme Court, in its decision of March 24th 1983 (reporting judge Vivas Marzal), considered the action a form of coercion. Typically, this is an action that is characterized as use of force (on things) in order to compel people to do something, in this case, to pay taxes. Considering the action a coercion, given the letter of the law, requires considering the use of force on things a form of violence.[23]

On July 20, 2020 (reporting judge Palomo del Arco) the Supreme Court decided that deleting from the cloud data pertaining to an ex-partner was punishable as coercion. The decision does not even mention or argue that there is violence in this case. Yet, punishing the action under the article on coercion, forces the interpretation of the action as a violent one.[24]

It is important to stress that independently of whether these actions are naturally seen as coercions, the issue here is that the law, as written, cannot consider them coercions, unless the term 'violence' applies to them.

The spiritualization of the notion of violence has been extended to other cases. On STS 17 March 2021 (reporting judge Magro Servet), the failure to pay pension or alimony (a crime that falls under article #227), is presented in the Supreme Court's argument as a form of 'economic violence'. It might eventually be argued that the crime in question falls under article #172 (on coercion) since the perpetrator of the offense prevents another person with *violence* from doing something not prohibited by law (such as providing necessary goods for their children, keeping them in the same school or having the same after-school activities), or compels them to do something they do not want to do (such as moving to a less expensive residence).[25]

These are all cases in which the domain of application of the term 'violence' has been expanded within the law. Are these instances of the phenomenon that we have characterized as semantic flexibility? That depends on whether the spiritualized use of 'violence' is seen as an alternative equilibrium accessible from the more restricted sense that originally operated in the law, a sense that applies 'violence' only to physical violence. Namely, it depends on whether the spiritualization of the term in

[23] The sentence can be accessed at https://vlex.es/vid/-76650807 (last accessed December 20, 2022).

[24] https://vlex.es/vid/847197267 (last accessed December 20, 2022).

[25] https://vlex.es/vid/863070312. See also https://noticias.juridicas.com/actualidad/noticias/16117-el-supremo-define-el-impago-de-una-pension-alimenticia-como-violencia-economica/ (last accessed December 20, 2022).

the interpretation of the law can be seen as continuous with past usage, something that the judges appear to assume. An alternative position is to argue that the term 'violence' is not flexible. That 'violence' always applied to the cases considered above and that we are slowly discovering the meaning of 'violence'. This would be to adopt a view similar to Mercier's as regards 'marriage' and to Perry's as regards 'cruel'.[26]

Are these cases instances of the phenomenon that we have characterized as semantic tolerance? That depends on whether the case can be interpreted as the creation of new uses of the term 'violence' that cohabitate even within the law, or as importing into the law uses that exist previously in the community. As we stated in section 1, the dividing lines between tolerance and flexibility are not entirely clear and distinct, even if tolerance and flexibility themselves are conceptually different.

Is the spiritualization of violence in Spanish law a case of conceptual engineering? We believe it is. It is clearly intentional, it alters some of the characteristic marks of a concept that was previously embedded in the law, and it has, as an effect, the extension of the domain of application of the term 'violence' for legal purposes.

Conceptual engineering is commonly tied to the amelioration of concepts, the project of revising and modifying concepts to make them more refined, more useful, more reflective of social reality and epistemically more fruitful.[27] A precise characterization of the notion of *amelioration* is difficult to come by, and there is quite a lot of discussion in the literature as regards what constitutes an amelioration,[28] but the idea of affording ourselves *better* concepts, and *better* tools to understand ourselves and our social environment is intimately tied to the project of amelioration. Now, is the spiritualization of violence brought about by the Spanish Supreme Court an amelioration? In a sense, yes, since it allows judges to acknowledge the severity of certain actions and punish them with what, from a common-sense perspective, is an appropriate penalty.

On the other hand, prolific engineering may have also undesirable consequences.[29] Given the spiritualization that the notion of violence has undergone, we may ask ourselves if cases of pacific resistance, such as sitting down to block the access of authorities to some buildings to carry out their orders, an action performed

---

[26] Adopting such a position would require regarding the articles of the law that distinguish violence from intimidation to be also in error, for intimidation should then be subsumed under violence (and that would arguably be a case of conceptual engineering). We thank Oliwia Przybyszewska for raising this point.

[27] As a referee has mentioned, there is a lively discussion on whether concepts really are the target of conceptual engineering. We do understand that there are very substantial questions, and a lively debate, as regards whether concepts, conceptions, terms, meanings, or social practices are the targets of conceptual engineering. But here we have taken it at face value that *conceptual* engineering targets, in principle, concepts.

[28] See, among others, Simion (2017). For a discussion of Simion's proposal see McKenna (2018), Podosky (2018), Isaac (2021) and Valero (2022).

[29] As Isaac et al. (2022) argue, extensive engineering may have unintended side-effects that may be pernicious. Marques (2020) has argued that some pernicious effects of engineering may well be not just unintended.

by Occupy movements in the US and Europe, could end up being characterized, according to the Spanish Penal Code, as violent actions.

The tendency to spiritualize may result in a trivialization of violence. If insults can end up being considered a form of violence (and it is difficult to see why they should not, given the arguments in the economic violence case, or in the deletion of cloud data case), there might be pernicious effects on vulnerable parties in cases of abuse and mistreatment. For instance, in a case of gender based violence, the conduct of a woman that insults repeatedly her partner might be assimilated to her husband's physical attack and so, the latter portrayed even as self-defense.[30]

These issues constitute some of the perils of extensive engineering. They are clearly, we surmise, not cases of amelioration, and they require a reflection in depth.

5.6 Conclusions

Both in cases of semantic tolerance and in cases of flexibility in the law we have tried to highlight the connections to conceptual engineering. When the law recognizes the presence of more than one use of a term in a community (semantic tolerance) we have concluded that the courts do not engage in engineering, although we have also suggested the conditions under which the creation of a new use of a term for legal purposes could be seen as engineering. On the other hand, cases of semantic flexibility, in which the courts or the legislatures decide to extend the domain of application of a term in one way or another are, arguably, instances of engineering, although some doubts remain in some cases (such as the same sex marriage case) as to whether the notion in question is itself changed in any way. Finally, we have examined what we regard as possible dangers of conceptual engineering in the law.

Our purpose has been to raise questions, rather than provide answers. But we hope that the questions and issues raised here will contribute to advance the discussion.

Acknowledgments We are extremely grateful to the three reviewers for their helpful comments. Versions of this paper were presented at the Arché Conceptual Engineering Seminar, at the 6th Parma Conference on Semantics and Pragmatics in Parma and at the Jagiellonian University in Krakow. We thank the audiences for their questions and comments. Work on this publication has been supported by the AGAUR of the Generalitat de Catalunya (2021-SGR-00276 and 2021-SGR-00923) and by the Spanish Ministry of Science, Innovation and Universities through the María de Maeztu Excellence grant CEX2021-001169-M (funded by MICIU/AEI/10.13039/501100011033), grant PID2023-150066NB-I00 (MICIU/EU) and grant PID2024-155879NA-I00 (MICIU/EU).

[30] This line of defense has been in fact tried, although so far without success. See the web of penal and family lawyer Felipe Mateo Bueno, where he insists on the psychological mistreatment that women subject men to: https://www.mateobuenoabogado.com/hombres-maltratados-una-realidad-silenciada/ (last accessed December 20, 2022).

References

Epstein, B. (2015). *What Alabama's Roy Moore Got Right*. https://opinionator.blogs.nytimes.com/2015/02/16/what-alabamas-roy-moore-gets-right/. Last accessed May 26 2023

Hutton, C. (2014). *Word meaning and legal interpretation. An introductory guide*. Palgrave MacMillan.

Isaac, M. G. (2021). Post-truth conceptual engineering. *Inquiry*. https://doi.org/10.1080/0020174X.2021.1887758

Isaac, M. G., Koch, S., & Nefdt, R. (2022). Conceptual engineering: A road map to practice. *Philosophy Compass*. https://doi.org/10.1111/phc3.12879

Jackman, H. (1999). We live forwards but understand backwards: Linguistic practices and future behavior. *Pacific Philosophical Quarterly, 80*, 157–177.

LaPorte, J. (2004). *Natural kinds and conceptual change*. Cambridge University Press.

Marques, T. (2020). Amelioration vs perversion. In T. Marques & A. Wikforss (Eds.), *Shifting concepts: The philosophy and psychology of conceptual variability* (pp. 260–284). Oxford University Press.

Martí, G., & Ramírez-Ludeña, L. (2016). Legal disagreements and new theories of reference. In A. Capone & F. Poggi (Eds.), *Pragmatics and law* (pp. 121–137). Springer.

Martí, G., & Ramírez-Ludeña, L. (2020). On whales and fish. Two models of interpretation. *Jurisprudence, 11*(1), 63–75.

Martí, G., & Ramírez-Ludeña, L. (2021). Tolerance, flexibility and the application of kind terms. *Synthese, 198*(Suppl 12), 2973–2986.

McKenna, R. (2018). No epistemic trouble for engineering 'woman'. *Logos and Episteme, 9*(3), 335–342.

Mercier, A. (2007). Meaning and necessity: Can semantics stop same-sex marriage? *Essays in Philosophy, 8*(1), Article 14.

Perry, J. (2011). Textualism and the discovery of rights. In A. Marmor & S. Soames (Eds.), *Philosophical foundations of language in the law* (pp. 105–129). Oxford University Press.

Phillips, I. (2014). Cetacean semantics. *Analysis, 74*(3), 379–382.

Podosky, P.-M. C. (2018). Ideology and normativity: constraints on conceptual engineering. *Inquiry*. https://doi.org/10.1080/0020174X.2018.1562374

Ragués i Vallès, R. (2003). ¿Coacciones sin violencia? Apuntes sobre el difícil encaje de la legalidad en un sistema funcional del Derecho Penal. In L.E. Montealegre Lynett (coord.): *El funcionalismo en derecho penal: Libro homenaje al profesor Günther Jakobs*, t. II: 481–497.

Ramírez-Ludeña, L. (2018). Interpretación evolutiva y tolerancia. Un análisis crítico de la Sentencia del Tribunal Constitucional Español sobre matrimonio entre personas del mismo sexo. *Discusiones, 22*, 153–188.

Ramírez-Ludeña, L. (2023). Statutory interpretation and binding precedents in the civil law tradition. In T. Endicott, H. Kristjansson, & S. Lewis (Eds.), *Philosophical foundations of precedent* (pp. 418–430). Oxford University Press.

Reuter, K. (2018). Dual character concepts. *Philosophy Compass, 14*(1). https://doi.org/10.1111/phc3.12557

Sabbarton-Leary, N. (2010). Descriptivist reference from metaphysical essence. *dialectica, 64*(3), 419–433.

Sawyer, S. (2020). The role of concepts in fixing language. *Canadian Journal of Philosophy, 50*(5), 555–565.

Scharp, K. (2013). *Replacing truth*. Oxford University Press.

Simion, M. (2017). The "Should" in conceptual engineering. *Inquiry, 61*(8), 914–928.

Tobia, K., Newman, G., & Knobe, J. (2020). Water is and is not H_2O. *Mind & Language, 35*, 183–208.

Valero, I. (2022). Ameliorating at the joints. A permissive normative framework for conceptual engineering. *Inquiry, 67*(10), 3728–3749. https://doi.org/10.1080/0020174X.2022.2063174

Wilson, M. (1982). Predicate meets property. *The Philosophical Review, 91*(4), 549–589.

Chapter 6
Labeling Unlabeled Identities

Katherine Ritchie

Abstract There is a longstanding debate about the harms and benefits of appealing to identities and using identity labels in social, political, and academic contexts. While there are potential harms with both labeling lacunae and with using labels, by considering the interplay of power, the interests and epistemic positions of the powerful and marginalized, and the role self-labeling plays, I argue that the introduction and usage of labels for both dominant and marginalized identities can bring important benefits and that strategies can be taken to mitigate harms. This shows, at a minimum, that blanket claims that identity terms are destructive or that identity labels do more bad than good are too strong.

> I'm a bitch, I'm a lover
> I'm a child, I'm a mother
> I'm a sinner, I'm a saint
> I do not feel ashamed
> —Meredith Brooks 'Bitch' from *Blurring the Edges* (Brooks, 1997)

The role identities and identity terms ought to play in policy and education is currently being debated in both academic circles and within the public sphere. Some argue that relying on identities and using identity labels leads to more harm than good. It divides when we need unity; it trains our attention on particularities, when universals ought to be emphasized (e.g., Lilla, 2017; Appiah, 2018). Advice suggesting that we ought to stop labeling to promote better self-esteem, teaching, and parenting abounds.[1] Further some argue that minimizing labeling can promote a more just world. For example, Leslie argues that reducing the use of labels may help limit prejudice and the extent to which people essentialize groups (2017: 418–

[1] A Google search found top results with advice to "stop labeling" or "ditch the labels" in *Psychology Today*, *Forbes*, and many sites with parenting and teaching advice.

K. Ritchie (✉)
University of California, Irvine, CA, USA
e-mail: ritchiek@uci.edu

M. G. Isaac et al. (eds.), *New Perspectives on Conceptual Engineering - Volume 3*, Synthese Library 515, https://doi.org/10.1007/978-3-031-98535-5_6

420). Others argue that identity talk is needed to understand the ways people are differentially positioned, to reckon with systematic forms of oppression, and to explain differences in lived experience and the distribution of resources (e.g., Alcoff, 2006; Young, 1990).

There are many ways to intervene in these debates. My focus here is on a question related to language. Think, for a moment, of the expressions that make up the lexicon of a particular language as tools. These tools facilitate discussion, explanation, making salient, and reference. In addition to considering a language as it currently is, we can also imagine alternative ways that a language could be. We can imagine a language with different tools, with more tools, or with fewer. For instance, we can envisage a language that does not have expressions that categorize or label people in ways related to race, gender, or sexual orientation. Or we could imagine one that is more expressively rich and allows for further identities to be picked out. Here I consider the harms and benefits of introducing new identity labels into a language. Considering these will help to shed some light on what identity terms can do and whether or when labels ought to be introduced or eliminated.

As I am using the term, *labels* are expressions that are lexicalized in a language (i.e., are in the lexicon of the language) and which purport to pick out (objects, events, kinds, ...) and categorize. When asking whether a label should be introduced for something, *A*, we are asking whether there are reasons to prefer a language with a compact lexicalized way of picking out *A* to a language without one.[2] We can also consider the effects of various sorts of labels. For instance, what differences are there between labeling with adjectives rather than nouns (Gelman, 2003; Markman, 1989; Ritchie, 2021b; Wierzbicka, 1986)? Are more fine- or coarse-grained labels preferable (for a particular purpose or in a particular context)? My aim here is not to answer the many important questions about what sorts of labels to use, but to consider the effects of labels as a broader class by contrasting languages with more limited linguistic resources and those with richer resources in order to argue that linguistic innovation regarding identity language can bring important benefits.

We have labels for many social identities.[3] People are labeled and self-label with *woman, Black, American, Republican, queer, liberal* and so on. There are also identities for which there are no labels whatsoever (in a given language, *L*). While I cannot populate a list of labels for identities that have none, *genderfluid, nonbinary, cis(gender)*, and *able-bodied* are identities for which labeling lacunae have recently been filled. There are also identities for which there are labels (in *L*) that are infrequently used. For example, in many contexts certain identities are taken to be the default or norm, so much so that labeling them might seem odd. For instance, in some contexts it is uncommon to use *white, heterosexual*, and *cis*.

[2] Both nouns and adjectives are labels on this construal given the understanding of lexical categories in Szabó (2015).

[3] In using *identity* here and throughout this paper I have in mind what are sometimes called "collective identities" like racial, gender, social-economic class, and religious identities. I do not mean to be focusing on what is sometimes called "individual identity" which is tied to a particular self-representation.

While my focus is on identities for which there are no labels (in *L*), in analyzing the dynamics of unlabeled identities, appeal to infrequently labeled identities will sometimes be useful in drawing out consequences of introducing and coming to use a label.

To answer the questions about whether and when introducing an identity label can be beneficial, I consider the effects of introducing and applying labels to unlabeled, but existent, identities.[4] I focus on these cases, given the following thinking. If labeling identities generally has negative effects that are not and cannot be outweighed by benefits, that strongly suggests that we should not *proliferate* identity terms. Perhaps given the way our language currently is we should not eliminate all the identity labels that already exist. It might not be possible to do so. It might lead to backlash that would nullify the benefits of a language without identity labels. But, if labeling identities is generally harmful, it is plausible to conclude that we should not add *more* identity labels.

Both sides in debates about identity labels are right in some respects—there are potential harms from using labels and from not using labels. I argue using two case studies that the introduction of labels for previously unlabeled dominant and marginalized identities can bring important epistemic, social, and political benefits and that potential harms can be mitigated. This points to conditions under which introducing a label is positive. It also shows that broad criticisms that labeling is more bad than good or that appealing to identities is always destructive are too strong. Returning to the tool metaphor, it can be useful to have a greater range of tools, but this does not show that any one tool (e.g., some particular label for an identity group) is the best expression to use for all purposes or in every context. Rather, just as a hammer can be useful for some purposes and a wrench can be better for others, identity labels can facilitate explanation, understanding, organizing, and other positive ends, but other sorts of expressions can be useful as well. This claim also allows us to recognize that labels can be used in ways that are harmful (just like hammers!). The call for richer representational resources for identities does not involve a mandate that we use identity talk in all contexts or for all (social, political, epistemic . . .) purposes.

My focus here is on language and on introducing expressions into a language. These projects fit into the broader project of conceptual engineering understood

[4] I focus on consequences here. I take these to be morally significant whether one is a consequentialist or non-consequentialist. In considering identity-talk, consequences are often brought to the fore. For instance, in arguing that identity talk is divisive, divisions have negative ramifications on society, and, therefore, identity talk should be limited or eliminated, one is relying on the consequences of identity talk. One might argue that agents have an interest in self-labeling and having the tools to do so is good in and of itself, without appeal to the consequences of labeling. I won't engage in that project here, but take it to be a welcome addition. I thank an anonymous referee for suggestions on this point.

as at least partially linguistic.[5] There is currently debate about what the target of conceptual engineering is (see Isaac (2021a, 2021b) for a useful taxonomy and arguments which I draw from here). Some argue its target is representational devices in general without requiring further specification (e.g., Burgess & Plunkett, 2013a, 2013b, 2020; Cantalamessa, 2019; Nado, 2019; Sterken, 2020). Others take its target to be something specifically mental (e.g., concepts; see Isaac, 2021a, 2021b). Still others take it to target something else, for instance something specifically linguistic (Cappelen, 2018). Since my focus is on language, whether this directly fits into a project of conceptual engineering depends on how one construes such projects. On a view on which engineering linguistic labels (i.e., introducing new words) is at least one way to engage in conceptual engineering, then my aims here do fall within conceptual engineering. On a view on which conceptual engineering is about concepts in particular, the project I am engaged in here is a form of what Isaac (2021a) suggests could be called *lexical engineering*.

It is worth noting that even if one construes conceptual engineering as targeting only concepts, one might still hold that lexical engineering is a means to changing concepts. For instance, one might hold that learning new words also involves acquiring or facilitating the use of a concept. I will not take a stand on questions about concept learning or acquisition or other questions about the how language affects cognition here. Instead, I will largely constrain my discussion to language, leaving what implications for concepts are to be drawn (or not) open depending on one's more general views on conceptual engineering and the mind-language interface.

The paper is structured as follows. I begin by arguing that there are unlabeled identities (*I*). While labels can make shared identities more salient, I argue that there is a meaningful notion of identity on which identities can exist without being labeled. Then I consider the effects of failures to label identities (*II*). I consider general effects that result from labeling lacunae as well as more specific effects, in particular harms, that come from failing to label dominant and marginalized identities. Next, I consider the effects that come from labeling, including worries that labels are divisive and that they promote bias and essentialism (*III*). I then turn to my main argument that labeling unlabeled dominant and marginalized identities brings important benefits (*IV*). Finally, I consider lingering worries and conclude (*V*). Labels and labeling can be positive; blanket claims that identity labels are harmful and should be avoided are too strong.

[5] For discussions and overviews of conceptual engineering see Cappelen (2018), Burgess and Plunkett (2013a, 2013b), Burgess et al. (2020), and Isaac et al. (2022).

6.1 The Existence of Unlabeled Identities

In order for the project of examining the effects of introducing labels for unlabeled identities to get off the ground, it must be possible for there to be unlabeled social identities. Whether there can be social identities for which there is no label whatsoever (in a linguistic community in which there are individuals with the identity) is controversial. For instance, Appiah (1996, 2005) argues that identities require labels.[6]

My aim here is not to argue for a complete metaphysics of identities, social groups, or other elements of the social world. Rather, my aim is to make a case that there can be unlabeled identities, by drawing on prominent views in social metaphysics. To begin to make this case, consider the fact that other social entities can exist without being labeled. For instance, Thomasson (2003a, 2003b) and Khalidi (2013) have influentially argued that recessions and racism could exist without our having any concepts or expressions that pick them out. Haslanger (2003: 314) argues that there can be social categories that are not named or explicitly classified in a language. She offers as a possible example children of widows (in a particular culture), which she suggests might involve a shared social position without a label for the category. This provides some prima facie evidence for unlabeled identities, insofar as other social entities are not inherently dependent on linguistic resources. To further bolster the case, let's consider a prominent way of understanding what identities are. (There may be other metaphysical views that also support the existence of unlabeled identities, but I will just focus on this one to offer a concrete illustration).

There is a robust tradition in feminist and critical race theory of understanding identities via an analogy with maps. As Cross argues, "one's identity is a maze or map that functions in a multitude of ways to guide and direct exchanges with one's social and material realities" (Cross Jr, 1991: 214; cited in Haslanger, 2005). Haslanger notes that while some use the metaphor of a social script, analogizing identities to maps "might be preferable insofar as it need not be understood linguistically, and may involve a 'map' of one's own body" (Haslanger, 2005: 283). The relevant social and material realities include the distribution of various resources (economic, political, epistemic, ...), the way social space is structured (e.g., in terms of bathrooms, grocery stores, elevators), and so on. Agents navigate these through combinations of norms, dispositions to act, ways one conceptualizes one's body, history, actions, and so on. Alcoff takes identities, which she says

[6] Dembroff and Saint-Croix (2019) argue that labels are not needed for identities. However, they hold that identities do require having a conceptual representation of a group as an identity requires "an ability to recognize the group and to recognize one's own relevant similarity with it" (Dembroff & Saint-Croix, 2019: 579). See Haslanger (2005) for arguments that Appiah's view of identity, and I take it Dembroff and Saint-Croix's as well, does not adequately account for people with white identities who do not take themselves to perform actions using a linguistic or conceptual representation that involves acting "as a White person" (2005: 277).

are "grounded in social locations" (2006: 9) to have "causal determinacy over our epistemic and political orientations to the world" (2006: 90).

Using the notion of a social map, Haslanger and Jenkins offer views of racial and gender identity, respectively, as follows:

> One has an X racial identity just in case their map is formed to guide someone marked as X through the social and material realties that are (in that context) characteristic of Xs as a group (Haslanger, 2005: 284)

> S has a gender identity of X iff S's internal 'map' is formed to guide someone classed as a member of X gender through the social or material realities that are, in that context, characteristic of Xs as a class (Jenkins, 2016: 414)

On these views, having an identity is understood in terms of maps that are tied to a social group or class. This condition is important, I take it, as racial and gender identities are meant to be social, shareable, and not wholly private or individualistic.[7] Social classes/groups can have multiple members. They are caused to be and constituted by diverse social factors, not just individualist features or feelings, beliefs, or representations. For instance, as Jenkins argues, there is an objective element to identities as "there must be some genuine [perhaps imperfect] correspondence between the norms people take to be relevant to themselves and the norms associated with the relevant gender class in at least some context" (2016: 412). By tying identities to social locations or classes/groups, identities meet these constraints.

Since, on this view, identities always involve a relation to a social class/group, without the existence of the requisite relata, there will be no possibility of having an identity that involves a map formed to guide one based on the social or material realities of the (non-existent!) social class/group. If a map view of identity is adopted and we are aiming to determine whether there can be unlabeled identities, we face two related questions: Can there be unlabeled social groups/classes? And, if so, can there be unlabeled identities? I take it that the answer to both is yes.

Groups like Haslanger's children of widows, as well as fine-grained intersectional groups for which we do not have compact labels, are examples of groups that seemingly exist without labels. Haslanger (2005) and Jenkins (2018) both consider cases of identities that do not appear to have come into existence with the creation of a label. Haslanger (2005) argues that multi-racial identities involve having an aggregate or "substantially fragmented" map. These maps guide people who have them in ways that cohere with someone marked as being in one race in some contexts or on some dimensions and in ways that cohere with someone marked as of another race in other contexts or on other dimensions. Jenkins argues that a genderfluid identity can be understood in terms of, e.g., a map that sometimes guides a person in ways that are characteristic of those classed as a woman and at other times guides them in ways characteristic of those classed as a man (2018: 735-6). While Haslanger and Jenkins do not discuss a lack of terminology and we now

[7] See Jenkins (2018) for an argument that relying on gender as class (or some other sort of gender roles that are not identical to gender identities) is also needed to avoid circularity.

have labels like *genderfluid* and *multiracial* it is compatible with their views that one could have a fragmented, aggregate, or fluid map without there being a label for a class or group of people the map is meant to guide.[8]

To put the argument more succinctly, groups, including those that are unlabeled, position people in ways that involve systematic applications of norms, constraints, enablements, and so on.[9] People positioned in these ways will plausibly come to internalize ways of moving through social reality that are sensitive to these norms, constraints, and so on. That is, they will plausibly come to have an identity tied to an unlabeled group even though they do not (and cannot!) self-label.

The ways one is guided by an identity might be opaque even to the person themself. This might even be the standard case.[10] Later, a person might interrogate their behavior or thought patterns—perhaps with the help of a therapist or close friend—and come to better understand how their identity was shaping the ways they felt, what options seemed available to them, how they interacted with others, and so on. This reveals a limitation with the map analogy, since with an actual map, one looks at it, interacts with it, and consciously uses it to guide their movement. In contrast, on the map view of identity, a gender map (or other identity map) can be internalized so that it guides a person without requiring them to have conscious access to it. It might function a bit more like a map in a self-driving car—guiding one through patterns of action, conceptualization, etc. in ways that do not involve one thinking "these are the sorts of actions people with identity *I* or in group *G* (ought to) take".

Before moving on, I want to note that I do not mean to imply that labels can have no effects on people's identities. For instance, I do not endorse the following as a complete picture of identity label introduction.

Stage 1: There are some people with identity, *I*, for which there is no label in a language *L*.

Stage 2: A label, *N*, is introduced into the lexicon of *L* allowing people to more easily share that they are *I* and their *I*-related experiences.

On this sort of picture, the process of introducing a label *only* changes our linguistic (and perhaps conceptual and epistemic) resources. This is far too simplistic. Introducing labels can have metaphysical effects on identities as well. That does not require that there was no identity to begin with. (For there to be something to

[8] While the account of an aggregate or fragmented map might be the right way to analyze some people's multiracial identities, others with multiracial identities might feel that there are contexts in which no racial norms seem to apply to them. Similarly, Jenkins argues that an agender identity involves not having a map that functions to guide one in ways that are characteristic of any gender class (Jenkins 2018: 736).

[9] Some take social categories (i.e., social groups or classes) to require representations. For instance, Mallon argues that representations of categories are "a crucial part of the mechanism by which" categories are constructed (Mallon 2016: 8). I do not take this to be a requirement on category/group/class construction.

[10] See Flores and Camp (forthcoming) for arguments that identities can be centered in one's ways of thinking about others and oneself without *beliefs* about them.

change, there must have been something there to change!) Rather, it involves the claim that labels can change identities. For instance, norms might be felt to apply more strongly and might be more strictly enforced. Hacking's notion of looping effects suggests one route by which this might occur (1995; 1999). In coming to be labeled or to self-label as an *N* the map that guides those who have identity *I* might be transformed. Appiah states that "labels operate to mold what we may call identification, the process through which individuals shape their projects—including their plans for their own lives and their conceptions of the good life—by reference to available labels, available identities" (2005: 66). While Appiah takes labeling to be a requirement for identity and I do not, I agree that labels can shape one's conceptions of oneself, the narratives one constructs, and so on.[11] To construe this within the map analogy, one might note that if a certain part of a mountain is labeled as the highest peak this might change the way people interact with the physical environment. For instance, hikers might seek a direct route to this point, thereby creating a path that will eventually come to be marked on subsequent maps. Someone labeling themself as an *N* might shape how they interact in the social world, how others interact with them, and so on.[12]

I have argued that there can be unlabeled identities by relying on the notion of identities as involving internalized maps. Given that there can be unlabeled identities, the question about what linguistic strategy ought to be adopted—to label or not to label—looms.[13] In the next two sections I consider the effects that can come from labeling and failing to label. I frame the discussion largely in terms of the harms that come with failures to label and the harms that come with labeling, although I recognize that several of the features cannot be neatly categorized as generally positive or negative. I return to considering the particularities of identities, power, and the value-laden effects of labels in 6.4.

[11] There is also evidence that the way one labels can affect how one conceptualizes oneself. For instance, Walton and Banaji (2004) found that when people were required to report preferences with noun constructions ("I am a coffee-drinker") rather than a verb phrase ("I drink coffee a lot") they took these preferences to be stronger, more stable, and more resilient. This suggests that self-labeling with nouns can shape self-perception.

[12] Thanks to Katharine Jenkins for suggesting this way of extending the analogy.

[13] Some (but not all) of the arguments I give in the remainder of the paper could be used to support a project that involves both a linguistic introduction and a metaphysical creation component. For instance, the creation of an identity and the introduction of a label for it might allow for people to come to have more authentic identities, which would be beneficial. I won't consider this variant further, but it is open to someone who disagrees with the metaphysical arguments given in this section to pursue some of the same strategies I do below.

6.2 Effects of Failures to Label

Failures to label mask the existence of identities and their corresponding social groups. It will emerge that some of the effects of labeling lacunae are general, while others vary with differences in power. In the case of dominant identities, masking serves to promote the interests of the group and to make their hegemony harder to see, theorize, and challenge. In the case of marginalized identities, masking is a form of hermeneutical injustice. It works to maintain marginalization and impedes forging bonds of solidarity. In the next three subsections I discuss general effects of the lack of labels and more specific effects of failing to label dominant and marginalized identities.

6.2.1 General Effects of a Lack of Labels

Shared terminology allows us to more easily collaborate and coordinate. Examples of this abound in every area of life. We use shared terminology in efforts to address climate change, to apply theories of differential equations to new domains, to coordinate to develop new technology, to differentiate and distinguish between styles of music, and to collaborate with a co-parent to raise children. Every field involves some terminology that allows for researchers, students, and others to more concisely express ideas that are then used in further theorizing and applications.

Shared terminology also allows us to offer explanations and justify or criticize the existence of descriptive and normative regularities. Nouns, one labeling device, have been argued to promote certain sorts of kind-based reasoning and explanation (Bigler & Liben, 2007; Gelman, 2003). Failing to have labels for social identities—either dominant or marginalized—can inhibit our ability to offer easily digestible explanations.

A lack of a compact label can impede explanation and prediction. For instance, consider a case in which A is asking for an explanation for why someone behaved in a particular way and B offers one of a variety of potential responses.

1. A: Why did Bradford do that?
   B: Because he's gay / Black / cis / white / a tall funny linguist.

While all of B's answers sound felicitous in some contexts, some of them seem suitable (as in, conversational participants would accept them as answering the question) in many contexts, while others are much more limited. Consider explanations in terms of the first two features. These could involve an appeal to stereotypes. For instance, one can easily imagine that B is homophobic or racist in appealing to Bradford being gay or Black as an explanation. However, taking being Black or gay to be explanatory need not involve derogation. For instance, suppose A is asking why Bradford put his hands high on the steering wheel while slowly and calmly asking the police officer who pulled him over whether he could reach into the glovebox for his registration. B responds: because he's Black. It is not just stereotypes about purportedly intrinsic biological features to which social explanations can appeal. Rather, social structure could also underpin why someone being Black, trans, and so on figure in explanations (Haslanger, 2016; Ritchie, 2019; Vasilyeva & Lombrozo, 2020). This makes social identities potentially explanatory in a wide range of contexts. It is more difficult to imagine a context in which a longer expression like "a tall funny linguist" would be explanatory.[14] In many contexts, B answering in that way would not be taken to be a suitable response to A's question. Whether there is a lexicalized label, either in the form of a noun or an adjective, makes a difference when offering explanations that are taken to be acceptable.

Expressive resources impact how well one can perform on tasks involving categorization and matching. For example, Lupyan and Zettersten (2021) found that shorter descriptions of a distinction (in terms of the number of words or number of clauses used in the description) were correlated with greater accuracy in drawing a distinction between two categories. Some of their stimuli involve Bongard problems and variants of them (Bongard, 1967). Bongard problems involve twelve geometrical representations in two sets of six. Participants are tasked with finding the difference between the images in the two sets; this is the solution to the problem.

For instance, in Example A in Fig. 6.1, the set of figures on the left are all triangular but differ in size, borders, and so on. The figures on the right are all circular but also differ along other dimensions. This problem was easy for participants to solve. They also had lexicalized expressions to easily draw out the difference—*triangle* and *circle*. In contrast, distinctions that were more difficult to compactly describe in a participant's language were harder to solve. For instance, while the difference in the sets in Example B can be described as being convex or concave, not all participants are familiar with these labels and far fewer solved this problem. Overall, Lupyan and Zettersten found that lower linguistic complexity was significantly correlated with successfully solving a problem and with greater accuracy in categorizing new exemplars into a category. These findings are intriguing but they are consistent with either (i) compact labels facilitating

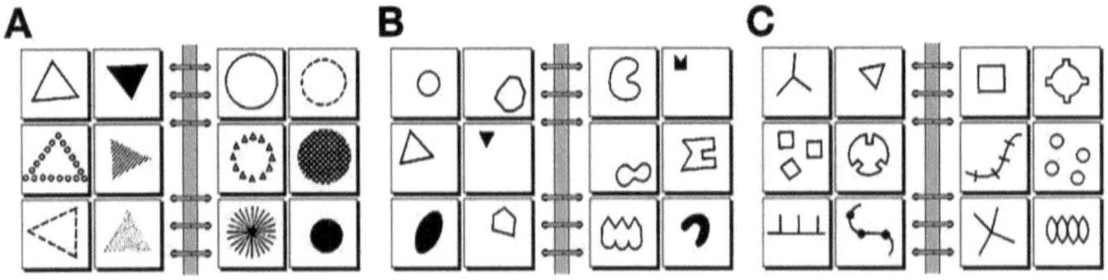

Fig. 6.1 Three examples of Bongard Problems (image from Lupyan & Zettersten, 2021)

[14] See Prasada and Dillingham (2006, 2009) and Haward et al. (2018) for empirical work showing that people take lexicalized expressions to be better explanations than complex noun phrases like 'tall funny linguist.'

categorization or (ii) certain categories being more natural for us to comprehend thereby both facilitating categorization judgments and explaining why the categories are named.

To differentiate between these Lupyan et al. (2018) tested several groups of participants with different access to formal education and language input (in the case of the latter, congenitally deaf Chinese children deprived of language input for much of their early lives). In a geometrical task requiring participants to find "the odd one out", they found that overt naming improved performance, that children with limited language input performed far worse than children with normal language input, and that verbal interference (repeating "a, b, c") selectively negatively affected participants responses on easy to name trials. These findings suggest that there is a connection between how easy it is for us to verbalize a category distinction and how well we can draw the distinction and correctly classify new instances. Labels appear to help facilitate our categorizing practices—for good or for ill. Next let's consider the effects of unlabeled dominant identities.

6.2.2 Specific Effects of Unlabeled Dominant Identities

When dominant groups go unlabeled their hegemony can be masked. Categories that are unlabeled are less salient and will be taken to be less explanatory. There is evidence that even when there are labels, not *using* a label can have unjust effects and, of course, a label cannot be used if it does not exist at all. For example, returning to example 1 above, in some contexts explanations that appeal to being white or cis will be interpreted as non-answers or as uncooperative.[15] By failing to have or use labels, the dominant group can normalize their interests, norms, practices, and so on as features of a neutral "non-identity" individualistic stance (Du Bois, 1924).

Sociolinguistic research has found that dominant groups in high school and university settings are often not labeled (Eckert & McConnell-Ginet, 1995; Kiesling, 2001). For instance, Eckert and McConnell-Ginet argue that "If the dominant category is not even labeled (and, as we noted earlier, in many schools it is not), then its distinctive interests are somewhat easier to ignore, its hegemonic control over social values and institutional norms more readily established. Two category labels in direct opposition reflect a live ongoing social struggle" (1995: 13). By failing to label themselves or be labeled by others the dominant group comes to enjoy an "unmarked default or 'normal' status within their school" (McConnell-Ginet, 2020: 71). Carrying this over to cases of unlabeled identities in a broader non-scholastic setting, McConnell-Ginet considers the introduction of the terms *cis*, *cisgender*, and *cishet.* She notes that "in many, many contexts, it still 'goes without

[15] Of course, these identities can figure in explanations and in some contexts they will be considered highly explanatory (see, e.g., discussions and memes using the label *yt* on Twitter). The label *yt* is pronounced like *white.*

saying' the people are cisgender" adding that cis people want to view themselves "as just 'normal' humans" (2020: 75).

Mills' notions of "an epistemology of ignorance" and "white ignorance" draw out a way those positioned as racially dominant can be ignorant of "the world they themselves have made" (1997: 18; see also Mills, 2007). He goes on to say that "Part of what it means to be constructed as "white" ..., part of what it requires to achieve Whiteness..., is a cognitive model that precludes self-transparency and genuine understanding of social realities" (Mills, 1997: 18). Other dominant identities, including those that are unlabeled, can also be masked to both in-group and out-group members. The failure to have or use labels for dominant groups allows for dominant identities to be obscured even more.

Masking and making dominant identities appear as the norm or as neutral non-identities, also makes individuals with non-dominant identities appear deviant. For example, consider accents. While these are not identities per se, they are clearly observable and can signal group membership or identity. Many take the Standard American accent (aka a broadcast English accent) to be an accent-less form of English. That is, they take it to involve no accent whatsoever and would not use a label like *Standard American accent*. On this conceptualization, other accents involve divergence from a norm or standard. They may be taken to involve mispronunciations, or to be funny, cute, or impossible to understand. What is taken to be standard is taken to be normal. And this normality is not just in terms of a (often false) view about statistical likelihood, but in terms of what is correct. On this view, it is not just that divergence from the norm is statistically odd, but that it is wrong; it is deviant in a morally-laden sense.

6.2.3 Specific Effects of Unlabeled Marginalized Identities

The lack of identity labels for marginalized groups can make it harder for members to make their experience and social situation intelligible, constituting a form of epistemic injustice. Let's begin by focusing on a broader failure with lacking conceptual or linguistic resources. In discussing nonbinary people, Dembroff argues that a conceptual lacuna "contributes to a hermeneutical injustice that arises from the failure to spread and charitably analyze the concepts and practices underlying nonbinary classifications" (Dembroff, 2020: 2). As Fricker defines it, hermeneutical injustice is "the injustice of having some significant area of one's social experience obscured from collective understanding owing to a structural prejudice in the collective hermeneutical resource" (Fricker, 2006: 100). A limitation in linguistic, conceptual, and epistemic resources affects what can be shared, discussed, and made intelligible (Dotson, 2012; Pohlhaus 2011; Fricker, 2006, 2007; Medina, 2012).

Limitations in linguistic, conceptual, and epistemic resources can render an aspect of someone unintelligible to everyone or to just some people. Clearly, if in a discourse context or socio-linguistic community there is no label for an identity it will be difficult, if not impossible, to communicate or share one's identity with others. As Mason (2011) and Dotson (2012) argue, this does not show that understanding or knowledge is impossible, full stop. Rather, they argue, even if the resources of the dominant discourse are limited, there can be alternative resources that surpass those used in dominant groups, and which allow for one with access to them to know and render intelligible their identities and experiences. In some cases, though, both dominant resources and alternative non-dominant hermeneutical resources can lack the tools to communicate about an identity. In this case, a person's identity will be difficult or impossible to interpret or understand for others and for themself.[16]

A lack of labels can also make organizing and fostering solidarity challenging. It will be more difficult to find commonalities and band together to work to mitigate oppression that comes from marginalization without linguistic resources for shared identities and shared forms of experience. Labels are one tool that facilitate, or perhaps better, serve as a precondition for organizing.

To write laws and advocate for policies to abate the marginalization and oppression faced by an identity group, lawmakers and politicians need to have access to linguistic resources to pick out the relevant group. For instance, in implementing affirmative action or instating reparations for past injustice conceptions of the groups that a law is aiming to serve are needed. As discussed in section 6.2.1 above, having a compact label, rather than a long description, is apt to mark out a category more clearly and allow it to figure in more digestible explanations. Linguistic resources of some sort are needed to advocate for new anti-oppression policy; labels help facilitate making these clear and compelling.

Finally, there can be metaphysical effects that come from a lack of resources to conceptualize and discuss identities. While I have argued that there can be unlabeled identities, the lack of resources to talk and think about them limits how people see others and how individuals conceptualize themselves. This could lead some to being seen as having or actually coming to have identities that are not authentic to themselves. For instance, Fricker suggests that "hermeneutical injustice can mean that someone is socially constituted as, and perhaps even caused to be, something they are not, and which it is against their interests to be seen to be . . . Hermeneutical injustice is an epistemic injustice with social constructive power" (2006: 107-8). On a similar note, Ásta states, "Sometimes there is no location in the current context that we identify with. And sometimes there are very few contexts with locations that we identify with. Our struggle then is to make it the case that there are more contexts in which there are locations that we can identify with" (2018: 123). There is a strong interpretation of both Fricker and Ásta on which they are saying that one cannot have an identity without it being labeled, conceptualized, or contextually

[16] I do not take this to require that all thinking is done in language. Rather, it is to say that without shared linguistic or conceptual resources for an identity, representing it at all, even if it is an aspect of one's own identity, can be difficult or impossible. There are interesting issues regarding how discursive and iconic representational formats relate to identities, that I cannot address here but I hope to in future work. See Flores and Camp (forth.) for discussion of centering identities with either images or linguistic representations.

recognized. Given my arguments in 6.1, I do not adopt this strong view. However, I agree that linguistic or conceptual limitations can elicit false or unjust perceptions of who or what a person is[17] and, perhaps even more troubling, could shape people in ways that give rise to dysphoria and feeling inauthentic.

6.3 Effects of Labeling

Some of the preceding discussion makes clear what introducing and using a label is apt to do. Introducing labels for both powerful and marginalized groups can serve to unmask. In the case of the powerful, this can lead to seeing what might be viewed simply as "the norm" as in the interest of dominant groups. In the case of marginalized groups, this can serve to make identities more salient and shareable, allowing for greater solidarity and organizing power. Insofar as labels facilitate drawing category distinctions, explaining, and categorizing members into kinds, introducing labels where there were none will make these tasks easier.

Using labels can also have harmful effects. It has been argued that labels divide, lead to cognitive and normative distortions, and increase bias, overgeneralization, and essentialism. As I stated above, much of the discussion in 6.2 focused on the harms of labeling lacunae. Much of the discussion in this section is focused on the harms of labeling.

Labels have been argued to be divisive. Sometimes these arguments come from those with power, for instance by white people wanting to focus less on race or cis people not wanting to be called *cis* and pathologizing being trans. As McConnell-Ginet nicely puts it "Those who *can* resist being marked often do so. They decry labels as 'divisive' and affirm people's individual distinctiveness. Treat us as individuals, they say. Don't lump us all together. Don't generalize" (2020: 76, *emphasis original*). McConnell-Ginet's words here echo arguments that have been given against identity politics and that are used to deny the relevance of race, gender, sexual orientation, and other identities. There is, of course, also something compelling about these sorts of arguments—labels certainly can facilitate overgeneralizing.

In psychology there is a large body of research on the effects of groups and group membership on performance, preferences, and judgments. For instance, people tend to have preferences for in-group members (Tajfel et al., 1971). Using generics and labels increases in-group preferences and leads to positive information being generalized more to in-group members while negative information is generalized

[17] Some metaphysical views of social categories require a bit more nuance on this claim. For instance, Ásta (2018) would reject the idea that a person with adequate standing in a context can falsely take someone to be a member of a social category. On her account if one is taken to have some identity (i.e., taken to be a member of some social category) by someone with adequate standing, one really is a member of that category. She would, I believe, take the phrasing in terms of unjust categorization to be acceptable.

more to out-group members (Bigler et al., 1997; Baron & Dunham, 2015; see Liberman et al., 2017 for an overview and additional citations). These effects have been found even in minimal groups like those based on matching shirts that are randomly assigned to participants. Introducing labels for identities that are more robust and not simply made-up in the setting of a psychology experiment is likely to have the same or stronger effects.

Research points in different directions as to whether labels are *required* for in-group and out-group effects (see Baron & Dunham, 2015; Baron et al., 2014 for evidence that labels are needed; see Jordan and Dunham (2021) for arguments that labels are not needed). Yet, regardless of the answer to this question, evidence shows that labels do augment in-group/out-group effects.

Labels have also been argued to promote psychological essentialism. Psychological essentialism is the view that we take some categories to be inductively rich, stable, and explanatory and to have an underlying unobservable essence that causes or otherwise realizes observable features (Gelman, 2003; Neufeld, 2022; Newman & Knobe, 2019). While not all labels appear to promote essentialism (e.g., nouns for artifacts like *hammer* do not, or do not to the same degree), some research finds that nouns for human groups like racial and gender terms do augment essentialist thinking. For instance, consider the generics in 2 and 3.

2. Asians are good at math.
3. Women are better caregivers than men.

The generics in 2 and 3 express stereotypes and may lead to strong judgments that there is a biological essence of the social kinds Asians and women that causes their members to share these features (Cimpian & Markman, 2011; Rhodes et al., 2012). Labels used outside of generics, as in 4, or in conjunction with images have also been argued to essentialize (e.g., Carnaghi et al., 2008; Gelman, 2003; Gelman & Heyman, 1999; Ritchie, 2021b).

4. Sara is a female / a queer.

This has led some to argue that we ought to avoid social generics or even social labels more generally (Haslanger, 2011; Leslie, 2017; Wodak & Leslie, 2017; but see Saul, 2017; Ritchie, 2019, 2021a).

Yet, other research suggests that labels for social identities do not always promote essentialism. They may, for instance, be accepted in virtue of social and contextual features that are not related to individual kind members having a biological essence (Noyes et al., 2021; Vasilyeva & Lombrozo, 2020). Moreover, while essentialism can have negative effects, it is not always correlated with increased prejudice and in some cases is even correlated with *reduced* prejudicial attitudes (Haslam et al., 2002; Peretz-Lange, 2021). Labels clearly have significant effects on social cognition, but what generalizations hold is a topic that is still being actively studied.

6.4 Introducing Labels for the Unlabeled

We have seen that there are harmful effects of labeling lacunae. There are also harms that can come with labeling. There are myriad complexities involving the introduction of labels. No simple prescription can adequately navigate these. Here I consider introducing labels for an unlabeled dominant identity and an unlabeled marginalized identity and argue that there are cases in which both introductions can bring benefits. Moreover, I argue that harms may not result, and steps can be taken to mitigate their possible occurrence. Using both examples illustrates the differences power plays in label introduction and shows that benefits are not intrinsically tied to the social standing of an identity group. Overall, I conclude that appeals to identities and the use of identity labels can be positive. Labels cannot be simply and summarily dismissed as harmful.

The introduction of labels for both unlabeled dominant and marginalized identities has positive epistemic effects. We saw that one primary harm of failing to have or use labels for dominant identities is that the group's hegemony and the particularity of their interests are masked. By labeling a previously unlabeled dominant identity, it becomes clearer that this is not a neutral identity. The label can be used to describe and explain, perhaps revealing to us another form of hierarchical social structure. If all identities—including dominant identities—are labeled, this can also help to mitigate the view that there is something deviant about non-dominant identities. In the cases I am focused on, labels are not creating new identities. Rather, introducing labels allows for the existence of identities and the power and normative dynamics they involve to be brought out into the open. This, in turn, could change the dynamics of power and identities themselves (perhaps through looping effects (Hacking, 1995, 1999)). Yet, even in these cases, an identity is not being created ex nihilo with the introduction of a label.

Introducing a label for a previously unlabeled marginalized identity, perhaps as a part of a broader "conceptual revolution" (Langton, 2010: 460), works to alleviate a hermeneutical injustice. Linguistic introduction and conceptual overhauling are strategies used to remedy the lack of resources available to make sense of and communicate identities. As Pohlhaus (2012: 719) explains:

> when there is a tension between the world of experience and the resources that we use to make sense of our experiences, for example when the proper language for describing an experience appears to be missing, or when our current concepts fail to track recurring patterns, we recalibrate our epistemic resources and/or create new ones until the tension between our resources and the experienced world is alleviated.

Allowing people to better understand, describe, and explain their identities and experiences are all positive effects of introducing a label for a previously unlabeled marginalized identity.

Expanding linguistic resources has epistemic benefits for anyone with access to the resources. The other benefits and harms of using labels vary depending on group status and whether the label is used for self/in-group members or for others/out-group members. While I argue that both introductions can bring important benefits that are not always outweighed by harms, it is important to note once again that I am not offering a simple prescription to always increase labels or labeling. There are clear cases in which identities are appealed to in harmful ways. I do not take this to be up for debate. Even the most committed proponent of identity politics or person who takes Black studies, feminist philosophy, queer studies, and so on to be vital academic projects will openly accept that there are some instances in which identities and identity labels are used in harmful ways. For instance, a use of a label might be intended to demean. I take it whole classes of expressions, like slurs, ought not be used (except perhaps in cases of appropriation).

The cases I consider here do not involve introducing new *derogatory* terms for out-group members. Often, but not always, derogatory terms are used to label groups that are already labeled. For instance, Tirrell (2012) discusses the introduction of new derogatory terms for Tutsi by Hutu preceding and during the Rwandan genocide. She argues that these were used amongst Hutu (e.g., by out-group members) to emphasize differences between groups that were already labeled in other ways, to dehumanize, and ultimately as one way to put in place norms that made murder acceptable. As I will argue, the epistemic access to the existence of an identity and whose interest it will be in to introduce new linguistic tools differ starkly between the cases of introducing a new label for an already labeled identity and introducing a label for a previously unlabeled identity. I make the case that introducing labels for the unlabeled can be beneficial by considering first dominant and then marginalized identity labels.

Consider an introduction of a label, *D*, for a previously unlabeled dominant identity. (I will work with the abstract *D* but it may be useful to think of *cis* as an example of a recently introduced label for a previously unlabeled dominant identity.) When out-group members begin to label the dominant identity using *D*, those with non-dominant identities can better describe and explain the features just discussed. The label can help in consciousness raising and in remedying epistemic occlusion. These are all benefits.

These beneficial effects are not limited to uses of *D* by out-group members. In-group/self-labeling with *D* can also involve recognizing, rather than taking for granted, that one has a dominant group identity. In this way, self/in-group labeling can also be positive.

Of course, not all instances of self/in-group labeling by people with a dominant identity are positive. Even well-intentioned people with dominant identities, can use self/in-group labeling in ways that continue to center powerful groups. For instance, think of people who frequently describe themselves as having white privilege. This can involve an apologetic stance or a feeling of self-hatred. For instance, in an anecdote I recently heard, someone said they hated their white skin so much that they claimed to want to tear it off. While recognizing privilege is a valuable step in undermining oppression, focusing on the privilege of those in power is not likely to have significant remedial effects on power imbalances, and may reinforce the power dynamics already in play.

There is a second sort of self/in-group labeling in the case of dominant identities that brings with it the potential for serious harms. As we saw, using labels can heighten inter-group effects, including increasing preferences for in-group members and judging that out-group members are more apt to do bad deeds than in-group members. If members with a dominant identity begin to use *D* to talk about themselves, this could lead to strengthened preference for other Ds. Uses like these are manifest in the language of dominant identity group movements—like white power movements. While clearly the language used in these movements is not the only factor that is relevant when considering the increased prejudice and violence they bring, it is one component.

The cases just considered show that in-group uses of labels by dominant group members have a range of effects. To show how these can range from positive to extremely negative, let's consider three members of a dominant identity group D who use *D* to label themselves and their in-group. One person strongly identifies as D, one is apologetic about their membership in D, and one recognizes their membership in D without strong identification or apology. Since the first strongly identifies as a D they are most likely to use *D* in ways that will promote further solidarity—*we Ds are the ones actually getting the short end of the stick!*—thereby continuing to advantage and bolster the extant dominance hierarchy. The apologetic figure will continue to center dominant identities in discourse in ways that sustain some features of hierarchies (DiAngelo, 2018). The third might reflectively engage in using *D* in positive ways—recognizing and expressing that there are structural forms of oppression, not denying their identity, and actively working to promote change. These cases suggest that for dominant group members, stronger self-identification with a label is apt to elicit more negative effects.

Next, let's consider who is likely to introduce a label for D. By being unlabeled, the dominant group continues to mask their hegemony and retain power. Insofar as it is in the interest of the dominant group to retain their dominance in this way, a label is likely to be introduced and used by non-group members first. In such cases, out-group uses are likely to be positive; they involve allowing for explanation, consciousness raising, and organizing that would be difficult or impossible without shared resources.

It is also relatively rare for dominant group members to self-label or use in-group labels even when labels exist (e.g., recall discussion of *cis* and *heterosexual*). So, it is possible there will be almost no in-group/self-labeling. If there is in-group/self-labeling some of it may also be genuinely positive (aimed at recognition and change). Building on this case, uses of *D* that are most likely to have harmful effects are those that involve reaction and a feeling of the loss of power and these uses may not develop at all. This shows one sort of case in which introducing a label for an unlabeled dominant group is apt to bring important benefits and suggests that these will not always be outweighed, given interests in not self-labeling.

Other cases are different. Dominant groups can also have interests in making their dominance manifest, particularly if they aim to solidify it further through an active movement to repress others in society. In this case the introduction of a label is unlikely to bring greater benefits than harms. In fact, it might be a tool primarily used to oppress.

Both sorts of cases are possibilities. My aim here is to show that given the interests dominant group members often have for avoiding labels for their identities (as discussed in 6.2.2), there is a class of cases in which labels will be introduced by out-group members with marginalized identities. These bring epistemic and social and political benefits. In-group/self-labeling may only rarely occur, and it too could be positive. There are good reasons to think that introducing a label *D* can bring real benefits.

Next, consider an introduction of a label, *M*, for a previously unlabeled marginalized identity group. (I will work with the abstract *M* in this case, but to make the case more concrete one might think of *non-binary* as an example of a recently introduced label for a previously unlabeled marginalized identity.) In-group uses of the label might support greater self-understanding and diminish feelings of dysphoria. It can also facilitate organizing and help to promote unity and cohesion. For instance, in considering the way in which labels are used in class struggle, Iris Marion Young states "One can and many do say, 'I am a worker,' as a badge of pride and identity. . . .when this happens . . . one has formed a *group* with other workers with whom one has established self-conscious bonds of solidarity" (Young, 1994: 727). In addition to being organizing tools, in-group uses of identity labels might also, as Young suggests, change or create new groups, which may become hubs for increased political power (see also, Ritchie, 2021c).

Out-group uses of *M* could also be in the service of greater recognition of M in ways that are positive. But not all out-group uses will be beneficial. They could involve mere symbolism, without doing anything to deconstruct hierarchies (Táíwò, 2022). They might also involve labeling in ways that increase stereotypes, out-group bias, and essentialism. These sorts of harms are connected to the extent to which an out-group label is used in a way that "otherizes" members of the group. The greater degree to which the use of an identity label seems to be saying that *they* have some feature in a way that contrasts with features *we* have, the higher the likelihood that it will have negative effects. The pattern here is the inverse of that which I argued holds for dominant self-labeling. In that case, the more a person self-identifies with the identity group, the higher the likelihood of negative effects of in-group labels, like *D*. Here, the more one represents a marginalized out-group as distinct and other, the more likely it is for uses of *M* to have detrimental social and psychological effects.

Given interests and epistemic access, new labels for unlabeled marginalized identities are likely to be introduced by those who have the marginalized identity. It is people with the identity who recognize that there is a, as Pohlhaus put it, "tension between our resources and the experienced world." Individuals with marginalized identities are also the group with the strongest interest in having tools to better conceptualize and discuss their identities. For these reasons, in-group usage and self-labeling are likely to precede out-group/other labeling. Further, as Mason (2011) and Dotson (2012) argue, new resources might be introduced as alternative resources or resources that are more widely available in dominant discourse. If the introduction is only in an alternative discourse that is shaped by those with the identity, it is likely to involve in-group and self-labeling as the discourse and resources are being shaped, at least in part, by those with the once unlabeled identity. These uses bring with them important positive effects—epistemic, social, psychological. Harms may not result, and steps can be taken to mitigate them. For instance, labels might be used by those within particular communities without aiming to introduce them into the wider dominant discourse. Strategies could be developed to work to avoid the appropriation of conceptual resources by powerful groups (Davis, 2018) or elite capture (Táíwò, 2020, 2022).

This shows that there is a class of cases in which the introduction of labels for marginalized identities is apt to bring benefits. Harms may come about as well, but the epistemic, political, social, and other benefits are real factors speaking in favor of augmenting linguistic resources.

6.5 Concluding Remarks

I've argued that uses of labels for both marginalized and dominant identities can have positive and negative effects. By considering the interplay of epistemic access and interests tied to power, I have argued that there are classes of cases in which introducing labels for unlabeled identities brings important (epistemic, social, political, cognitive, ...) benefits and that given interests and alternative strategies, harms can be mitigated. The social world and the way we think and talk about it is extraordinarily complex. I do not mean to neglect this. Rather, my aim here was to push back on the anti-identity and anti-identity language strand in some popular and academic discourse. Identity labels can bring real benefits that are not always outweighed by harms. A more nuanced position recognizing that even adding more identity labels can bring real benefits is needed.

I close by considering three final worries. First, someone might argue that while perhaps not all labels do more harm than good, the potential for harm from a label is astronomical, suggesting that we ought to reduce labeling with extant terminology and avoid introducing any new labels. To harken back once more to the tool metaphor, the argument could be framed in terms of taking identity and other social labels to be more like guns (or nuclear weapons) than hammers. The claim is then that labels have the potential to be so detrimental that they ought not be available to people at all. I have argued against this extremely strong view by considering the epistemic, psychological, social, and political benefits of labels and labeling.[18]

Second, one might object that I have failed to consider the longer-term effects of introducing labels for unlabeled identities. And, the argument might go, once a broader time scale is considered it will be found that labels are more harmful than beneficial. As a generalization, this is an extremely strong thesis. The ways that racial, sexual orientation, and gender labels have figured in successful struggles for

[18] I thank an anonymous referee for suggesting this concern and way of extending the tool analogy.

rights and protections provide reason to think it is too strong and that it should be rejected. However, it might still be pressed, and rightly so, that there are longer-term harms that need to be considered.

To make the case in a vivid way, let's focus on the introduction of a label for a previously unlabeled marginalized identity *M*. Suppose that while there are some initial positive effects from in-group and self-labeling with *M*, after some time the label is used by those with dominant identities to stereotype people who are M, promote bias against Ms, and ultimately to carry out atrocities that include murdering many people who are M because they are M. Clearly this is a heinous scenario. But what we should conclude about the introduction of the label is, I think, not obvious.

The label plays a role in the story, but it is not the only factor. No one would argue that the introduction of the label was *the* cause of genocide.[19] The power dynamics that existed below the surface of the linguistically available resources were there before. In fact, the relationship between the introduction of the label and the genocide is plausibly highly sensitive to many changes in background conditions, making it a less appealing causal explanation (Lewis, 1986; Woodward, 2010; Woodward, 2021). But, given what I have argued here, the introduction of the label did contribute to bringing these to the surface.

Would it have been better if the label were never introduced? Certainly it would be better if the genocide had not occurred, but precisely what the role of a label for Ms played in the story is not easy to pin down. There are possible scenarios in which a label was introduced, it led to benefits like Ms better understanding their identities and banding together to improve their standing in society, and it did not lead to the atrocities in this imagined scenario. I take cases like these to illustrate once again the intricate and complex connections between language, identity, and power that have and ought to be further explored, but which a single paper cannot fully address. My claim is quite humble—there are cases in which introducing labels for unlabeled identities is beneficial. These ought to be recognized and reveal that any general claim that identity labels are bad and should be avoided is too strong.

Finally, one might argue that it would be better for there to be no identity labels at all. This argument could be made in a more or less idealized way. As a point about what linguistic resources we ought to have given the social world as we find it, I take this line of thought to be unconvincing. Here's an argument to show why. I take it there are identities (i.e., they exist). Further, it is better to have linguistic resources to talk about entities (very broadly construed) that exist than failing to have such resources. So, it is better for us to have linguistic resources to talk about identities (i.e., identity labels) than not, given the actual ontology of social reality.[20] It may be

[19] See Tirrell (2012) for discussion of the ways language is one piece of how the Rwandan genocide came to be. She emphasizes, citing Semelin, that the role of language and propaganda should not be overemphasized as "there is nothing to prove that this, on its own, leads to the unleashing of a massacre" (Semelin, 2007: 199).

[20] This is not to say that it is better to have derogatory labels, of course. But it is a very strong and controversial view that all identity labels are like slurs, for instance.

better to live in a world without any identities (although I do not really know what such a world would be like). If so, it *might* be better to have no identity labels. One might argue that in such a world labels would not be *tracking* anything in reality and would only *create* divisions.[21] Here I did not consider this (fictional) scenario, but rather asked about the benefits and harms of introducing labels in the social world as we presently find it. If the ontology of the social world were radically different, the effects of identity labels and whether they ought to be in a language at all might also differ.

My aim here was to investigate a small part of the complex dynamics of language, power, knowledge, and identity by considering the introduction of identity labels. I argued that the introduction and usage of labels for existent but previously unlabeled dominant and marginalized identities brings (epistemic, social, political, cognitive, . . .) benefits. Even given the complexity and particularities found across situations and contexts, there are reasons in favor of augmenting a linguistic toolkit to include further identity labels.

Acknowledgements Previous versions of this paper were presented at the University of California, Irvine Identities and Epistemic Injustice Workshop, 2021 Social Metaphysics Conference, the Vermont Mind, Language, and Social Change Workshop, the Pacific APA, the Conceptual Engineering Online Seminar, the Words Workshop, and at colloquia at the University of Vienna (in the Vienna Colloquia in Analytic Philosophy series), the University of Illinois—Urbana Champaign, and USC. I thank these audiences for their helpful feedback. In particular, I thank Lauren Abruzzo, Anna Boncompagni, Jeremy Evans, Carolina Flores, Katharine Jenkins, Zoë Johnson King, Alex Madva, Rebecca Mason, Eliot Michaelson, Bernhard Nickel, Daniel Wodak, 2 editors and an anonymous referee for suggestions and comments that significantly improved the paper.

References

Alcoff, L. M. (2006). *Visible identities: Race, gender, and the self.* Oxford University Press.

Appiah, K. A. (1996). Race, culture, identity: Misunderstood connections. *The Tanner Lectures on Human Values, 17*, 51–136.

Appiah, K. A. (2005). *The ethics of identity*. Princeton University Press.

Appiah, K. A. (2018). *The lies that bind: rethinking identity, creed, country, color, class, culture.* New York: Liveright Publishing Corporation.

Ásta. (2018). *Categories we live by: The construction of sex, gender, race, and other social categories by.* Oxford University Press.

Baron, A. S., & Dunham, Y. (2015). Representing 'Us' and 'Them': Building blocks of intergroup cognition. *Journal of Cognition and Development, 16*(5), 780–801.

Baron, A. S., Dunham, Y., Banaji, M., & Carey, S. (2014). Constraints on the acquisition of social category concepts. *Journal of Cognition and Development, 15*(2), 238–268.

Bigler, R. S., Jones, L. C., & Lobliner, D. B. (1997). Social categorization and the formation intergroup attitudes in children. *Child Development, 68*(3), 530–543.

[21] Although there are reasons to think that there are benefits to labeling in promoting agency, see Camp and Flores (2024).

Bigler, R. S., & Liben, L. S. (2007). Developmental intergroup theory: Explaining and Reducing Children's Social Stereotyping and Prejudice. *Current Directions in Psychological Science., 16*(3), 162–166.

Bongard, M. M. (1967). *Pattern recognition*. Hayden Book Co., Spartan Books.

Brooks, M. (1997). *Blurring the edges*. Capitol Records.

Burgess, A., Cappelen, H., & Plunkett, D. (2020). *Conceptual engineering and conceptual ethics*. Oxford University Press.

Burgess, A., & Plunkett, D. (2013a). Conceptual ethics I. *Philosophy Compass, 8*(12), 1091–1101.

Burgess, A., & Plunkett, D. (2013b). Conceptual Ethics II. *Philosophy Compass, 8*(12), 1102–1110.

Burgess, A., & Plunkett, D. (2020). On the relation between conceptual engineering and conceptual ethics. *Ratio, 33*(4), 281–294.

Camp, E. & Flores, C. (2024). Playing with labels: Identity terms as tools for building agency. *The Philosophical Quarterly, 74*(4), 1103–1136. https://doi.org/10.1093/pq/pqae092

Cantalamessa, E. A. (2019). Disability studies, conceptual engineering, and conceptual activism. *Inquiry: An Interdisciplinary Journal of Philosophy*, 1–31.

Cappelen, H. (2018). *Fixing language: An essay on conceptual engineering*. Oxford University Press.

Carnaghi, A., Maass, A., Gresta, S., Bianchi, M., Cadinu, M., & Arcuri, L. (2008). Nomina sunt omina: On the inductive potential of nouns and adjectives in person perception. *Journal of personality and social psychology, 94*(5), 839–859. https://doi.org/10.1037/0022-3514.94.5.839

Cimpian, A., & Markman, E. M. (2011). The generic / non-generic distinction influences how children interpret new information about social others. *Child Development, 82*, 471–492.

Cross, W. E., Jr. (1991). *Shades of black: Diversity in African-American identity*. Temple University Press.

Davis, E. (2018). On epistemic appropriation. *Ethics, 128*(4), 702–727.

Dembroff, R. (2020). Beyond binary: Genderqueer as critical gender kind. *Philosophers' Imprint, 20*(9), 1–23.

Dembroff, R., & Saint-Croix, C. (2019). 'Yep, I'm Gay': Understanding agential identity. *Ergo: An Open Access Journal of Philosophy, 6*, 571–599.

DiAngelo, R. (2018). *White fragility: Why it's so hard for white people to talk about racism*. Beacon Press.

Dotson, K. (2012). A cautionary tale: On limiting epistemic oppression. *Frontiers: A Journal of Women Studies, 33*(1), 24–47.

Du Bois, W. E. B. (1924). *The gift of black folk: The negroes in the making of America*. Washington Square Press.

Eckert, P., & McConnell-Ginet, S. (1995). Constructing meaning, constructing selves: Snapshots of language, gender and class from Belton High. In K. Hall & M. Buchholtz (Eds.), *Gender articulated: Arrangements of language and the socially constructed self* (pp. 469–507). Routledge.

Flores, C., & Camp, E. (forthcoming). "That's all you really are": Centering social identities without essentialist beliefs. In S. Haslanger, et al. (Eds.), *Mind, language, and social hierarchy*. Oxford University Press.

Fricker, M. (2006). Powerlessness and social interpretation. *Episteme, 3*(1-2), 96–108.

Fricker, M. (2007). *Epistemic injustice: Power and the ethics of knowing*. Oxford University Press.

Gelman, S. A. (2003). *The essential child: Origins of essentialism in everyday thought*. Oxford University Press.

Gelman, S. A., & Heyman, G. D. (1999). Carrot-eaters and creature-believers: The effects of lexicalization on children's inferences about social categories. *Psychological Science, 10*(6), 489–493.

Hacking, I. (1995). The looping effects of human kinds. In D. Sperber, D. Premack, & A. J. Premack (Eds.), *Causal cognition: A multidisciplinary approach*. Clarendon Press.

Hacking, I. (1999). *The social construction of what?* Harvard University Press.

Haslam, N., Rothschild, L., & Ernst, D. (2002). Are essentialist beliefs associated with prejudice? *British Journal of Social Psychology, 21*, 87–100.
Haslanger, S. (2003). Social construction: The “Debunking” project. In F. Schmitt (Ed.), *Socializing metaphysics* (pp. 301–325). Rowman & Littlefield.
Haslanger, S. (2005). You mixed? Racial identity without racial biology. In S. Haslanger & C. Witt (Eds.), *Adoption matters: Philosophical and feminist essays*. Cornell University Press.
Haslanger, S. (2011). Ideology, generics, and common ground. In C. Witt (Ed.), *Feminist metaphysics* (pp. 179–207). Springer Verlag.
Haslanger, S. (2016). What is a (social) structural explanation? *Philosophical Studies, 173*(1), 113–130.
Haward, P., Wagner, L., Carey, S., & Prasada, S. (2018). The development of principled connections and kind representations. *Cognition, 176*, 255–268.
Isaac, M. G. (2021a). What should conceptual engineering be all about? *Philosophia Philosophical Quarterly of Israel, 5*, 1–13.
Isaac, M. G. (2021b). Broad-spectrum conceptual engineering. *Ratio: An International Journal for Analytic Philosophy, 34*(4), 286–302.
Isaac, M. G., Koch, S., & Nefdt, R. (2022). Conceptual engineering: A road map to practice. *Philosophy Compass, 17*(10), e12879., 1–15. https://doi.org/10.1111/phc3.12879
Jenkins, K. (2016). Amelioration and Inclusion: Gender identity and the concept of woman. *Ethics, 126*(2), 394–421.
Jenkins, K. (2018). Toward an account of gender identity. *Ergo: An Open Access Journal of Philosophy, 5*.
Jordan, A., & Dunham, Y. (2021). Are category labels primary children use similarities to reason about social groups. *Developmental Science, 24*(1), 1–13.
Khalidi, M. A. (2013). Three kinds of social kinds. *Philosophy and Phenomenological Research, 90*(1), 96–112.
Kiesling, S. (2001). Stances of whiteness and hegemony in fraternity men’s discourse. *Journal of Linguistic Anthropology, 11*(1), 101–115.
Langton, R. (2010). Review of epistemic injustice: Power and the ethics of knowing by Miranda Fricker. *Hypatia, 24*(2), 459–464.
Leslie, S.-J. (2017). The original sin of cognition: Fear prejudice, and generalization. *Journal of Philosophy, 114*(8), 393–421.
Lewis, D. (1986). Postscripts to ‘Causation’. In *Philosophical Papers Vol. II*. Oxford University Press.
Liberman, Z., Woodward, A. L., & Kinzler, K. D. (2017). The origins of social categorization. *Trends in Cognitive Sciences, 21*(7), 556–568.
Lilla, M. (2017). *The Once and Future Liberal: After Identity Politics*. New York: HarperCollins Publishers.
Lupyan, G., Wendorf, A., Rojas-Berscia, L. M., & Paul, J. (2018). Core Knowledge or language-augmented cognition? The case of geometric reasoning. *Proceedings of the 12th International Conference on the Evolution of Language (Evolang12)*.
Lupyan, G., & Zettersten, M. (2021). Does vocabulary help structure the mind? In M. Sera & M. Koenig (Eds.). *Minnesota symposia on child psychology: Human communication: Origins, mechanisms, and functions* (Vol. 40, pp. 160–199). John Wiley & Sons.
Mallon, R. (2016). *The construction of human kinds*. Oxford University Press.
Markman, E. (1989). *Categorization and naming in children: Problems of induction*. MIT Press.
Mason, R. (2011). Two kinds of unknowing. *Hypatia, 26*(2), 294–307.
McConnell-Ginet, S. (2020). *Words matter: Meaning and power*. Cambridge University Press.
Medina, J. (2012). *The epistemology of resistance: Gender and racial oppression, epistemic injustice, and resistant imaginations*. Oxford University Press.
Mills, C. (1997). *The racial contract*. Cornell University Press.
Mills, C. (2007). White ignorance. In S. Sullivan & N. Tuana (Eds.), *Race and epistemologies of ignorance* (pp. 11–38). State University of New York Press.
Nado, J. (2019). Conceptual engineering, truth, and efficacy. *Synthese*, 1–21.

Neufeld, E. (2022). Psychological essentialism and the structure of concepts. *Philosophy Compass, 17*(5).
Newman, G., & Knobe, J. (2019). The essence of essentialism. *Mind and language, 34*(5), 585–605.
Noyes, A., Dunham, Y., Keil, F. C., & Ritchie, K. (2021). Evidence for multiple sources of inductive potential: Occupations and their relations to social institutions. *Cognitive Psychology, 130*(130), 101422. https://doi.org/10.1016/j.cogpsych.2021.101422
Peretz-Lange, R. (2021). Why does social essentialism sometimes promote, and other times mitigate, prejudice development? A causal discounting perspective. *Cognitive Development, 59*, 101085.
Pohlhaus, G. (2012). Relational knowing and epistemic injustice: Toward a theory of willful hermeneutical ignorance. *Hypatia, 27*(4), 715–735.
Prasada, S., & Dillingham, E. (2006). Principled and statistical connections in common sense conception. *Cognition, 99*(1).
Prasada, S., & Dillingham, E. (2009). Representation of principled connections: A window onto the formal aspect of common sense conception. *Cognitive Science, 33*.
Rhodes, M., Leslie, S. J., & Tworek, C. M. (2012). Cultural transmission of social essentialism. *Proceedings of the National Academy of Sciences, 109*(34), 13526–13531.
Ritchie, K. (2019). Should we use racial and gender generics? *Thought: A Journal of Philosophy, 8*(1), 33–41.
Ritchie, K. (2021a). Essentializing language and the prospects for ameliorative projects. *Ethics, 131*(3), 460–488.
Ritchie, K. (2021b). Essentializing inferences. *Mind & Language, 36*(4), 570–591.
Ritchie, K. (2021c). Does identity politics reinforce oppression? *Philosophers' Imprint, 21*(4), 1–15.
Saul, J. (2017). Are generics especially pernicious? *Inquiry: An Interdisciplinary Journal of Philosophy*, 1–18.
Semelin, J. (2007). *Purify and Destroy: The Political Uses of Massacre and Genocide*. New York: Columbia University Press.
Sterken, R. K. (2020). Linguistic intervention and transformative communicative disruptions. In A. Burgess, H. Cappelen, & D. Plunkett (Eds.), *Conceptual engineering and conceptual ethics* (pp. 417–434). Oxford University Press.
Szabó, Z. G. (2015). Major parts of speech. *Erkenntnis, 80*(1), 3–29.
Táíwò, O. (2020). Identity politics and elite capture. *Boston Review*. http://bostonreview.net/race/olufemi-o-taiwo-identity-politics-and-elite-capture
Táíwò, O. (2022). *Elite capture: How the powerful took over identity politics (and everything else)*. Pluto Press.
Tajfel, H., Billig, M. G., Bundy, R. P., & Flament, C. (1971). Social categorization and intergroup behaviour. *European Journal of Social Psychology, 1*(2), 149–178.
Thomasson, A. (2003a). Foundations for a social ontology. *Protosociology, 18–19*, 269–290.
Thomasson, A. (2003b). Realism and human kinds. *Philosophy and Phenomenological Research, 67*, 580–609.
Tirrell, L. (2012). Genocidal language games. In I. Maitra & M. K. McGowan (Eds.), *Speech and harm: Controversies over free speech* (pp. 174–221). Oxford University Press.
Vasilyeva, N., & Lombrozo, T. (2020). Structural thinking about social categories: Evidence from formal explanations, generics, and generalization. *Cognition, 204*, 1–14.
Walton, G. M., & Banaji, M. R. (2004). Being what you say: The effect of essentialist linguistic labels on preferences. *Social Cognition, 22*(2), 193–213.
Wierzbicka, A. (1986). What's in a noun? (Or: How do nouns differ in meaning from adjectives?). *Studies in Language, 10*, 353–389.
Wodak, D., & Leslie, S.-J. (2017). The mark of the plural: Generic generalizations and race. In P. C. Taylor, L. M. Alcoff, & L. Anderson (Eds.), (pp. 277–289). Routledge.
Woodward, J. (2010). Causation in biology: Stability, specificity, and the choice of levels of explanation. *Biology and Philosophy, 25*(3), 287–318.

Woodward, J. (2021). *Causation with a human face: Normative theory and descriptive psychology*. Oxford University Press.
Young, I. M. (1990). *Justice and the politics of difference*. Princeton University Press.
Young, I. M. (1994). Gender as seriality: Thinking about women as a social collective. *Signs, 19*(3), 713–738.

Chapter 7
Conspiracy Theories Are Not Theories: Time To Rename Conspiracy Theories

Kevin Reuter and Lucien Baumgartner

Abstract This paper presents the results of two corpus studies investigating the discourse surrounding conspiracy theories and genuine theories. The results of these studies suggest that conspiracy theories lack the epistemic and scientific standing characteristic of theories more generally. Instead, our findings indicate that conspiracy theories are spread in a manner that resembles the dissemination of rumors and falsehoods. Based on these empirical results, we argue that it is time for both re-engineering *conspiracy theory* and for relabeling 'conspiracy theory'. We propose relabelling conspiracy theories as either 'conspiracy stories' or 'conspiracy narratives' to more accurately convey their true nature.

7.1 Introduction and Theoretical Background

Two claims are frequently made about conspiracy theories. First, conspiracy theories are argued to be *about conspiracies* (see, e.g., Basham & Dentith, 2016; Cassam, 2019; Coady, 2003; Cohnitz, 2018; Feldman, 2011; Harris, 2018; Keeley, 1999; Pigden, 2007; Räikkä, 2018). Second, conspiracy theories are argued to be *theories* (see, e.g., Butter, 2022; Dentith 2022, Duetz 2022, Hepfer 2015, Pigden 2007). Both claims suggest themselves by a compositional analysis of the term 'conspiracy theory'. Consider, for example, the composite terms 'music theory', and 'string theory'. Very roughly, music theory is a theory about music, and string theory is a theory about strings. Thus, without evidence to the contrary, we might simply

K. Reuter (✉)
University of Gothenburg, Gothenburg, Sweden
e-mail: kevin.reuter@gu.se

L. Baumgartner
University of Zurich, Zurich, Switzerland
e-mail: lucien.baumgartner@philos.uzh.ch

M. G. Isaac et al. (eds.), *New Perspectives on Conceptual Engineering - Volume 3*,
Synthese Library 515, https://doi.org/10.1007/978-3-031-98535-5_7

assume the minimalist account to be correct, according to which conspiracy theories are theories about conspiracies.[1]

Empirical evidence against the first characteristic has recently been put forward by Napolitano and Reuter (2021). The results of their experiments reveal a double dissociation of *conspiracy* and *conspiracy theory*: Not only are people willing to call a highly speculative and unsubstantiated claim a conspiracy theory even though no conspiracy is part of the claim, they also show that even if a conspiracy is part of the view that is put forward, laypeople are not inclined to call the view a conspiracy theory if the conspiracy has truly taken place.[2] What about the second pillar of the minimalist account: are conspiracy theories really theories? Very recently, several papers have raised doubts about the status of conspiracy theories as theories (Đorić, 2020; Frindte & Ina Frindte, 2023; Huneman & Marion Vorms, 2018; Napolitano, 2021). While Napolitano (2021) argues against the claim that conspiracy theories are theories but rather are self-insulated beliefs in conspiracies, others have defended the minimalist account (Butter, 2022; Dentith, 2022; Duetz, 2022, 2023; Pigden, 2007). Most of thc arguments in this debate are based on *individual* scholars' intuitions about *individual* conspiracy theories. What is missing, so far, is a comprehensive analysis of the term 'conspiracy theory' that transcends the scope of individual conspiracy theories and circumvents the reliance on individual scholars' intuitions.

This paper utilizes corpus-linguistic tools to provide empirical evidence challenging the categorization of conspiracy theories as genuine theories. Corpus analysis involves using computational techniques to systematically and quantitatively study extensive text data. This method is well-suited for investigating the general use of expressions.[3] Through a thorough examination of the usage patterns surrounding the term 'conspiracy theory' within both pre-built and custom-made corpora, we conduct a comparative analysis that juxtaposes conspiracy theories against both scientific as well as non-scientific theories.

Before attempting to determine whether conspiracy theories are truly theories, it is imperative to first address the question of what constitutes a theory. Regrettably, there is limited agreement regarding this matter. While sociologist Abend (2008) recognizes seven distinct meanings of the term 'theory', Duetz (2023) considers

[1] While the meaning of composite terms is often made up of the meanings of its parts, that is not always the case, e.g., the 'rainbow press' is not the press about rainbows.

[2] Napolitano and Reuter (2021) also raise the question of whether the concept of *conspiracy theory* encodes an evaluative dimension and present empirical studies that strongly suggest a positive answer. Particularists are particularly skeptical of attempts to define or re-engineer *conspiracy theory* as an evaluative concept (e.g., Dentith 2019 and Pigden 2007, see also Tsapos (2023) for a discussion on the notion of conspiracy theorist). Shields (2023) has recently argued that "we should be very circumspect regarding engineering concepts that are primarily sites of conceptual domination." (2023, p. 465) Napolitano and Reuter (2023) defend the re-engineering of *conspiracy theory* against the charge of furthering conceptual domination.

[3] Reuter and Lucien Baumgartner (2024), as well as Alfano (2024) provide introductory chapters to corpus analysis from a philosophical perspective.

three different senses of the term: established account, hypothesis, and hunch. Meanwhile, philosophers often differentiate between the syntactic view, which conceives of theories as a collection of theorems formulated in languages of predicate logic (Carnap, 1966; Hempel, 1966; Winther, 2020), and the semantic view, which equates theories with a set of models (Suppes, 1960; van Fraassen, 1989). Natural scientists rarely provide a definition of a theory, but instead emphasize the unique epistemic status of theories (National Academy of Sciences, 1998, see also Popper, 1963/2014). They assert that theories undergo testing, confirmation, falsification, substantiation, refinement, and revision in response to the observations and experiments of the phenomena that the theories are meant to explain. In this paper, we adopt the latter *pragmatic* approach to examine the status of conspiracy theories. Although this pragmatic approach is certainly inspired by a scientific perspective, we will see that less scientific uses of 'theory', e.g., 'fan theory' or 'my/your/her theory', satisfy these epistemic conditions for being theories to a very large extent. Consequently, if it is found that people do not test, confirm, substantiate, refine, and revise conspiracy theories, this would provide substantial evidence in support of the claim that conspiracy theories are not theories in any relevant sense of the term.

Our motivation for choosing a pragmatic approach is threefold. First, there is no definition of 'theory' along the syntactic or semantic lines, that commands widespread acceptance in the literature. In contrast, there will be few scholars who deny that people test, confirm, substantiate, refine, and revise theories. Second, as elucidated by Abend (2008) and Duetz (2023) in their discussions on the various conceptions of theory, different notions of theory come in different strengths. By adopting a pragmatic approach, we are not compelled to establish a definitive threshold beyond which it becomes implausible to consider something as a theory. Instead, our pragmatic framework permits us to evaluate the status of conspiracy theories along a continuous spectrum, as we substantiate in the subsequent analysis.

Third, and perhaps most importantly, only the pragmatic approach can be easily operationalized for empirical studies. We aim to examine whether conspiracy theories are regarded as theories through a linguistic analysis. If people engage in activities such as testing, confirming, and refining (conspiracy) theories, we can reasonably anticipate that they would also discuss these activities in talking and writing. Hence, examining the language used to describe the handling of theories and conspiracy theories can offer insight into their epistemic standing. To investigate this, we will observe the language employed by individuals to describe their own and others' interactions with theories. This method is comparatively straightforward as we only need to analyze the language rather than the many ways in which people use and engage with theories.

In the next two sections, we present our corpus-analytic results. Corpus Study 1 demonstrates that conspiracy theories not only fail to be subject of scientific and educational activities, it also reveals a frequent occurrence of spreading terminology preceding the term 'conspiracy theories', which we further investigate in Corpus Study 2. We subsequently find remarkable similarities between the way we talk about conspiracy theories and the way we discuss falsehoods and misinformation.

In the following two theoretical sections, we discuss possible objections and consequences of the empirical results. More specifically, we will make two proposals. First, we believe our results require us to *re-engineer* the concept of conspiracy theory. Both the minimalist account as well as some popular characterizations define conspiracy theories as theories. However, such definitions lack empirical justification when considering the actual usage and application of the term ‘conspiracy theory’. Second, we will argue that we should re-label conspiracy theories in a way that better reflects that conspiracy theories are not theories. In this regard, we propose the adoption of terms such as ‘conspiracy story’ or ‘conspiracy narrative’, which better encapsulate the true nature and characteristics of conspiracy theories.

7.2 Corpus Study 1

In order to gain insight into the epistemic standing of conspiracy theories, we propose to examine the language employed in characterizing the activities associated with scientific theories, non-scientific theories and conspiracy theories. A fruitful avenue for investigating the discourse surrounding these concepts involves the compilation of a substantial corpus of phrases following the pattern of “VERB [target term]”, e.g., “test theories”, “share conspiracy theories”, etc. Importantly, we need to compare the verbs preceding a whole range of different theories in order to paint a fairly accurate and representative picture. This is what we have done in Corpus Study 1.[4]

Verbs occur in many different positions in English sentences. For our purposes, we only observe verbs that occur directly in front of the terms of interest. Let us illustrate the main idea by using a pre-built and freely available corpus like the NOW corpus.[5] The advantage of using such a corpus is that readers can—after registering—easily replicate the results for themselves. In order to get the respective data, we enter <VERB theories> into the search field of the NOW corpus, then click on options and group by lemma in order to catch all different forms of the verbs. Table 7.1 (right hand side) shows the most frequent verbs preceding the term ‘theories’. This list of the 10 most common verbs provides some positive evidence

[4] A comparative corpus analysis is, of course, not restricted to verbs only. Further evidence for what theories are and how people specify theories can be collected by investigating adjectives that occur before the target terms: “ADJ [target term]”. Previous studies by Napolitano and Reuter (2021) and Reuter and Lucien Baumgartner (2024) have used a similar design, focusing on adjectives rather than verbs (see also Baumgartner, 2022; Meylan & Reuter, 2025; Reuter, 2011; Reuter et al., 2023; Sytsma et al., 2019; Willemsen et al., 2021). Their studies indicate that the term ‘conspiracy theories’, in contrast to the term ‘theories’, is primarily an evaluative term. While these corpus analyses suggest important differences between conspiracy theories and other theories, they do not provide strong enough evidence about the status of conspiracy theories as theories.

[5] The NOW corpus (Davies, 2016) can be easily accessed through this website: https://www.english-corpora.org/.

Table 7.1 A list of the 10 most frequent verbs in front of 'conspiracy theories' and 'theories'

Conspiracy theories		Theories	
Term	Number	Term	Number
Promote	885	Fan	1157
Spread	611	Test	320
Push	490	Develop	182
Believe	441	Discuss	125
Peddle	389	Offer	105
Embrace	243	Share	98
Share	151	Promote	96
Amplify	141	Support	96
Debunk	136	Debunk	95
Espouse	135	Learn	83

that a linguistic approach is likely to deliver some promising results. Among the most frequent terms we find 'test', 'develop', and 'support', which highlight some of the scientific activities. Other terms like 'discuss', 'offer', and 'learn', more strongly emphasize some of the educational aspects surrounding theories.[6]

Next, we enter <VERB conspiracy theories> into the NOW search field and observe the most common verbs occurring before 'conspiracy theories' (see Table 7.1 left hand side). A look at the ten most common verbs preceding 'conspiracy theories' reveals a highly frequent use of verbs referring to the spreading of information. 6 out of 10 verbs in the top 10 belong to that category. In contrast, no terms appear that clearly belong to the scientific or educational realm with the possible exception of 'debunk', which is also the ninth most frequent term in front of 'theories'.

While these results indicate some important differences between the way people talk about theories and conspiracy theories, we need to be careful in not overinterpreting those results. First, we have only focused on the 10 most frequent verbs. Second, the NOW corpus only consists of texts from news websites. As such it might not give a representative picture of the way ordinary people talk about theories and conspiracy theories. Third, a more comprehensive comparative analysis should also include several control conditions. Given these limitations, we therefore decided to build our own corpus, include a greater list of verbs, and to throw several control conditions into the mix.

7.2.1 Methods and Data

As control conditions, we decided to analyze the following target terms: 'critical race theory', 'fan theories', 'game theory', 'music theory', 'string theory', and

[6] The most frequent term 'fan' is wrongly identified as a verb by the NOW corpus. Note also that the NOW corpus is a dynamic corpus. Thus, numbers will change given that the corpus grows continously.

just 'theories'. The terms 'game theory', 'music theory' and 'string theory' were selected (a) because they are frequently discussed on the internet and would likely deliver a sufficiently large sample, and (b) because they are representative and much-discussed theories from three different domains. The terms 'critical race theory' and 'fan theories' were selected because the status of critical race theory and fan theories as theories is perhaps a little more controversial. Especially fan theories do not have any strong claim to being scientific theories. As such, we included control terms that also represent less demanding uses of the term 'theory'. If we were to find that a linguistic analysis of 'conspiracy theories' is similar to an analysis of 'critical race theory' and 'fan theories', then this would certainly affect the conclusions we can draw from that data.

For this study, we focus on data from the social media platform Reddit. Reddit is often referred to as "the cesspool of the internet," and it is well known to host a myriad of conspiracy theories. However, one can also find elaborate discussions on more academic topics, such as music theory or string theory. Previous research has shown that Reddit plays an important role in the diffusion of conspiracy theories from QAnon to anti-vaxxers (e.g., Cinelli et al., 2022; Engel et al., 2022; Shahsavari et al., 2020). Hence, Reddit provides abundant corpus data relevant to our investigation.

The data for this study consists of 12,973 target structures extracted from Reddit comments spanning from October 1, 2021, to December 31, 2021, which were collected using the Pushshift API (Baumgartner et al., 2020). The data was cleaned and syntactically annotated (PoS-tagging) in order to extract our target structures.[7] Except for modal verbs and participles (e.g., "competing theory"), verbs often do not directly precede singular direct and indirect objects without the addition of an article (e.g., "I read a theory"). Hence, the fact that we are focusing on constructions of the form "VERB [target term]" means that we are forced to use either the plural form or standing terms as our targets. We further excluded past, gerund, and present participle constructions, as well as modal verbs.

7.2.2 *Verbs and Categories*

In order to provide a quantifiable analysis of the verbs in our corpus, we need to group them into categories. This can be be done manually or using unsupervised, automated methods. One of the potential drawbacks of unsupervised methods is the limited interpretability of results. For this corpus study, we thus rely on manual

[7] The syntactic annotation was performed using `spacyr` (v1.2.1) in `R` (v4.1.0). Data cleaning includes addressing various elements such as special characters, inconsistent whitespace, fragmented text, and contractions, among other issues.

coding. We examine the 30 most frequent verbs for all seven target terms and categorize them into five distinct categories:[8]

(i) **scientific**: build, confirm, create, develop, discover, disprove, elaborate, falsify, form, prove, solve, test, write.
(ii) **educational**: cover, describe, discuss, explain, grasp, learn, study, teach, understand, visualize.
(iii) **consumptive**: hear, read, see.
(iv) **attitudinal**: accept, appreciate, believe, embrace, enjoy, hate, like, love, oppose, reject, support.
(v) **spreading**: fuel, peddle, post, promote, propagate, push, share, spew, spout, spread.

Whether or not terms belong to the category consumptive, attitudinal, and spreading, should be fairly uncontroversial, the categories scientific and educational are certainly less clear. We asked five independent coders to tell us whether these terms belong to the scientific or the educational sector. At least four out of five coders agreed on all terms except two: 'define' and 'know' were subsequently excluded from the analysis.

7.2.3 *Results*

Table 7.2 displays the percentages of the five categories among all terms selected for classification. Crucial for an evaluation of conspiracy theories are those terms that belong to the scientific and educational category. Figure 7.1 illustrates the proportions of educational verbs versus scientific verbs. As we can see, the term 'conspiracy theories' has the lowest proportions for both verb classes (5.34% educational, 6.73% scientific) taken together. For 'theories', we find the highest proportion of scientific verbs (33.04%), but also slightly lower numbers of educational verbs (18.74%). 'Fan theories' is located between the two. On the other hand, 'music theory', 'game theory', and 'critical race theory' are predominantly preceded by educational verbs (>50%) and have very low proportions of scientific verbs (<5%). Lastly, 'string theory' is the only target phrase diametrically opposed to 'conspiracy theory'. If we compare the (summed) proportions of educational and scientific verbs to spreading verbs (Fig. 7.2), we see that 'conspiracy theory' is an outlier, as it is mainly used with spreading verbs (52.06%).

[8] Our decision to focus on the 30 most frequent verbs in our analysis might seem arbitrary, yet it was informed by several pragmatic considerations. First, we observed a notable drop in the occurrence of verbs ranked 30th to 40th when juxtaposed with our target terms, which implies that the top 30 verbs could likely serve as a robust representation. Second, expanding our scope beyond the 30th rank, for instance to the 50th, significantly increased the variety of verbs encountered, complicating the analysis. Third, beyond the range of 30, we encountered increasingly general verbs like 'have' and 'get', which proved challenging to categorize without additional context.

Table 7.2 Shares of verb class per target phrase [%]

	Attitudinal	Consumptive	Educational	Scientific	Spreading
Conspiracy theories	29.70	6.17	5.34	6.73	52.06
Critical race theory	25.92	4.58	60.10	2.54	6.86
Fan theories	31.40	30.23	9.30	19.77	9.30
Game theory	25.00	7.93	64.02	2.44	0.61
Music theory	5.51	0.82	93.08	0.24	0.35
String theory	12.07	4.31	62.93	20.69	0.00
Theories	16.65	21.09	18.74	33.04	10.48

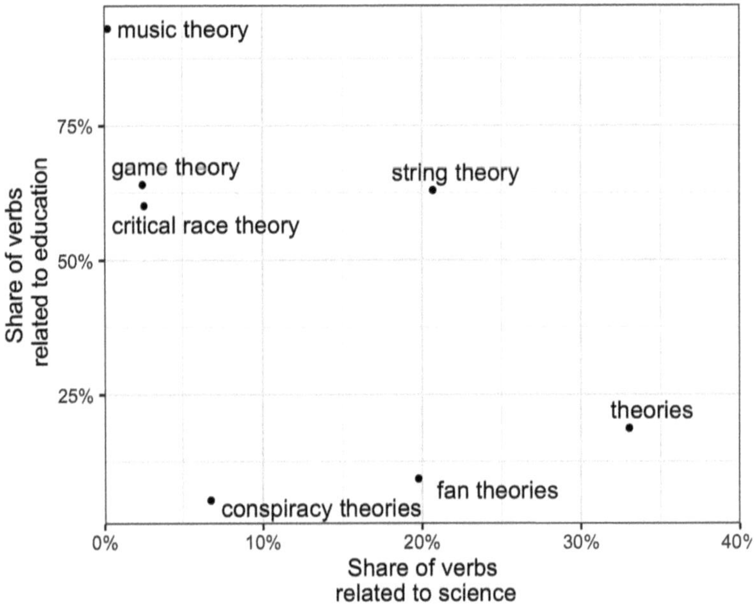

Fig. 7.1 Relation between the seven different terms on a two-dimensional space spawned by the components scientific and educational

7.2.4 *Discussion*

In our Corpus Study 1, conspiracy theories have revealed themselves to be markedly different from other theories. People do not write about conspiracy theories in the same scientific and educational manner in which they write about other theories: We neither find verbs that indicate people's scientific engagement with conspiracy theories, nor are conspiracy theories taught and studied like other theories.[9]

[9] Critical race theory, music theory, and game theory also have very few scientific verbs preceding them. Perhaps this indicates their status as theories in the social sciences in contrast to theories in the natural sciences.

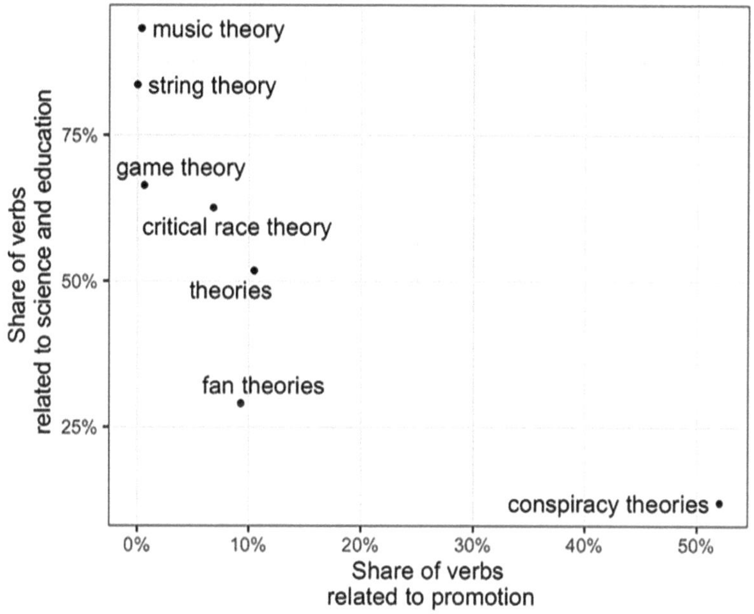

Fig. 7.2 Relation between the seven different terms on a two-dimensional space spawned by the components promotion/spreading and scientific+educational

At this stage, one might want to raise an objection against drawing too strong a conclusion from these results. After all, conspiracy theories are quite different to theories such as string theory and game theory. However, our control terms not only included clearly scientific examples but also non-scientific theories like fan theories. Fan theories are speculative interpretations created by fans to explain elements within various books or films. Yet, our comparative analysis reveals that educational and scientific terminology is far more frequently associated with fan theories—double and triple the rate, respectively—compared to conspiracy theories. We will revisit this objection in Section 4. At this juncture, the evidence suggests that conspiracy theories may share the nominal label of 'theories' but lack the substantial epistemic underpinnings of what one typically considers a theory.

7.3 Corpus Study 2

Not only did we observe a lack of scientific and educational verbs preceding 'conspiracy theories', we also found a class of verbs—spreading verbs—that frequently occur before conspiracy theories but hardly at all with any other of the tested theories. Why would people so frequently talk about conspiracy theories in that way? An answer might be provided by looking at other things that are spread, pushed and promoted.

Table 7.3 A list of the 6 most frequent nouns occuring after 'spreading', and 'peddling'

Spreading		Peddling	
Term	Number	Term	Number
Misinformation	7162	Lies	733
Awareness	5825	Drugs	557
Rumours	3873	Falsehood	314
Lies	3521	Conspiracy [theories]	228
Coronavirus	2818	Misinformation	200
Disinformation	2433	Rumours	191

Table 7.4 A list of the 10 most frequent verbs in front of 'conspiracy theories', and 'falsehoods'

Conspiracy theories		Falsehoods	
Term	Number	Term	Number
Promote	885	Spread	748
Spread	611	Perpetuate	370
Push	490	Peddle	304
Believe	441	Contain	142
Peddle	389	Promote	136
Embrace	243	Publish	121
Share	151	Propagate	116
Amplify	141	Tell	113
Debunk	136	Use	100
Espouse	135	Debunk	92

In order to explore phenomena that are spread and peddled, we enter <spreading NOUN> and <peddling NOUN> into the search field of the NOW corpus. The most common nouns are displayed in Table 7.3. These include 'misinformation', 'rumours', 'lies', and 'falsehood'. The term 'conspiracy theories' is the fourth most common noun appearing after 'peddling', the third most frequent noun after 'spouting' and the 21st most common noun after 'spreading'. In other words, many people seem to treat conspiracy theories on par with falsehoods, misinformation and rumours.

In order to further inquire into the similarities and dissimilarities of conspiracy theories on the one hand, and falsehoods and misinformation on the other, we decided to run a second corpus analysis in which we compare the categories of verbs preceding 'conspiracy theories' with verbs occurring before terms such as 'falsehoods'. In Table 7.4, we list the 10 most frequent verbs for 'conspiracy theories' and 'falsehoods' from the NOW corpus.

A direct comparison between the ten most frequent verbs before 'conspiracy theories' and 'falsehoods' displays remarkable similarities: The verbs 'promote', 'spread', and 'peddle' are among the top 5 most frequent verbs before both target terms. This warrants a more comprehensive comparative analysis. We therefore conducted a second corpus analysis using Reddit comments. Just like in our initial study, we selected Reddit for its rich repository of conversational data that aligns with our research interests. Furthermore, the platform's informal discourse offers a refreshing juxtaposition to the more structured texts found in the NOW corpus.

For this follow-up, we were interested in how 'conspiracy theories' (along with 'theories') align with broader categories of speech often associated with spreading verbs, viz. 'truths/falsehoods', 'information/misinformation', 'lies', and 'rumors'. We chose these terms based on the data presented in Table 7.3. We added antonyms for 'falsehoods' and 'misinformation' but not for 'lies' and 'rumors' as their antonyms are already partially covered by 'truth'. The antonyms are intended as an additional control condition and also act as a sanity check. The Reddit data for Study 2 consists of 11,997 new comments (spanning from October 1, 2021, to December 31, 2021), in addition to the observations for 'conspiracy theories' and 'theories' previously used in Study 1. The pre-processing and annotation used in Study 2 is identical to the one in Study 1.

7.3.1 Results

What we find is that 'theories' and 'truths' behave similarly, whereas 'conspiracy theories' is much closer to 'falsehoods' and 'rumors'. Figure 7.3 shows the proportion of verbs related to education versus science. As we can see, 'conspiracy theories' clusters nicely with 'falsehoods', 'lies', 'misinformation', and 'rumors',

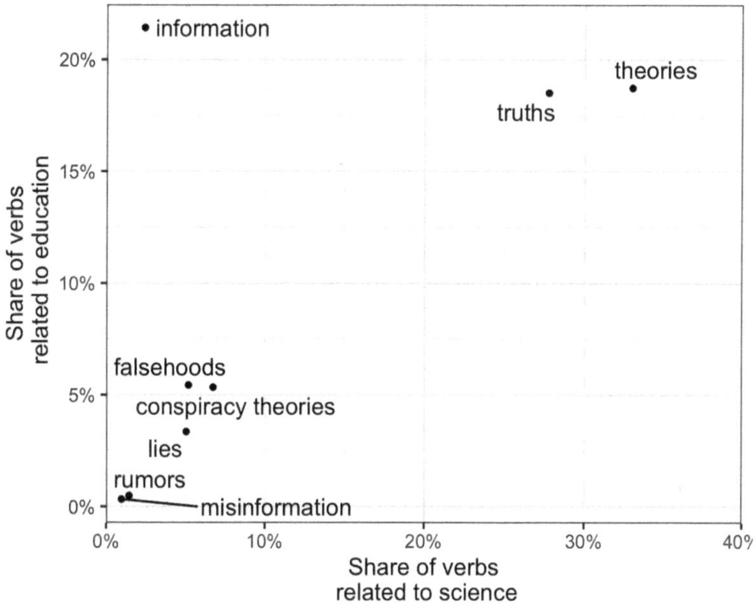

Fig. 7.3 Relation between the eight different terms on a two-dimensional space spawned by the components scientific and educational

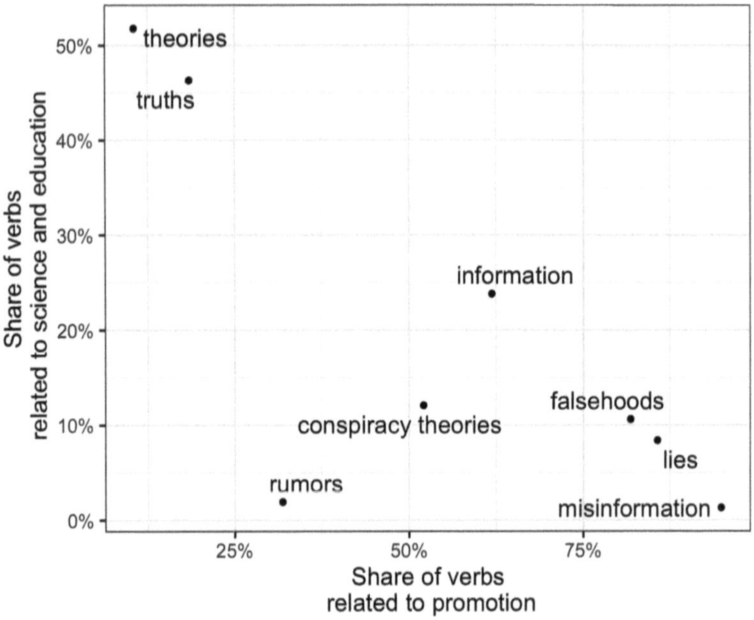

Fig. 7.4 Relation between the eight different terms on a two-dimensional space spawned by the components spreading/promotion and scientific+educational

which all have very low proportions of either verb category. In contrast, the share of educational verbs is quite high for 'information' (21.43%), 'theories' (18.74%), and 'truths' (16.66%). However, unlike 'theories' (33.04%) and 'truths' (27.78%), 'information' (2.38%) has a very low share of scientific verbs, similar to 'conspiracy theories' (5.34%). Figure 7.4 compares the proportions of verbs related to science and education with those related to spreading. Here, again, we see 'theories' and 'truths' are used very differently from all other target expressions. Their share of spreading verbs is a lot lower, and their share of scientific and educational verbs is a lot higher compared to the majority of the other terms.

7.3.2 Discussion

The results of Study 2 indicate that the term 'conspiracy theory' behaves in a manner similar to expressions associated with the dissemination of false information, while the word 'theory' does not. Analysis of verbs used in reference to conspiracy theories revealed a prominence of terms related to spreading, suggesting that people tend to focus on the dissemination aspect of conspiracy theories, similarly to misinformation and lies. These findings provide further support for the argument that conspiracy theories are not widely perceived as theories.

7.4 Summary and Objections

The results of Study 1 reveal a significant discrepancy in the manner in which we discuss conspiracy theories as opposed to other theories. While theories are commonly described as being tested, developed, and studied, conspiracy theories are very rarely discussed in these terms. Additionally, we discovered that, in contrast to theories, people often express that they spread, promote, and peddle conspiracy theories. Further examination of the prevalent use of spreading terminology in Study 2 shows a significant overlap between the way in which we discuss conspiracy theories and falsehoods. This leads to the conclusion that conspiracy theories are not, in fact, theories, and should therefore not be defined as theories. This conclusion may be met with resistance. In the following, we will discuss four objections that are likely to be raised against our claims.

Some may raise concerns that the Reddit comments that make up our main corpus may not be representative and therefore, the data collected could potentially present a distorted perspective. However, if this were the case, one would expect to observe vastly different results when analyzing the NOW corpus. While the list of verbs used in relation to conspiracy theories may not fully align between the Reddit and NOW corpus, the overall conclusions remain consistent: conspiracy theories are rarely preceded by scientific or educational verbs and instead frequently co-occur with terms related to spreading.[10]

Second, a counterargument could be advanced that the term 'theory' is a family-resemblance or exemplar concept, similar to the concept of *game* (Wittgenstein, 1953/2010; Rosch, 2024). While the activities associated with vastly different games such as poker and football are likely to be highly disparate, we still categorize both as games. Perhaps conspiracy theories and other theories are also only loosely connected. Some might even go as far as suggesting that the term 'theory' is polysemous (see also our discussion of the work of Abend (2008) and Duetz (2023) in the introduction): In the realm of science, a theory comprises a set of principles that shape and direct scientific inquiry. Contrastingly, in everyday language, 'theory' often simply denotes an idea that might explain a particular circumstance. While we do not disagree here with such possible distinctions, the important question to be raised is a different one: Do people really treat these theories differently in ways that are relevant given our pragmatic approach? Our results about fan theories suggest a negative answer. Fan theories are not spread and peddled but also tested and disproven (albeit perhaps to a sligthly lesser degree).

At this stage, one might argue that even fan theories possess a considerable level of sophistication. Instead, consider the example, "My theory for why my dog barks at the postman is his fear of bearded men." In such a case, don't we use "theory" more clearly in the sense of opinion that is not straightforwardly subject to being tested? The empirical data paints a different picture. As demonstrated

[10] We do not, of course, deny the possibility of variations in language usage regarding conspiracy theories among specific communities.

Table 7.5 A list of the 5 most frequent verbs in front of 'my theory', 'your theory', and 'her theory' on the NOW corpus

My theory		Your theory		Her theory	
Term	Number	Term	Number	Term	Number
Test	88	Support	35	Prove	62
Prove	80	Test	19	Test	62
Confirm	65	Prove	14	Share	59
Support	55	Disprove	14	Explain	50
Contradict	23	Think	14	Support	33

Table 7.6 A list of the 5 most frequent verbs in front of 'the conspiracy theory', and 'QAnon theories'

The conspiracy theory		QAnon theories	
Term	Number	Term	Number
Promote	155	Promote	12
Support	126	Push	7
Push	117	Spread	5
Spread	87	Retweet	3
Believe	65	Post	3

in Table 7.5, when we refer to "my theory", "your theory", or "her theory", the associated activities predominantly and clearly have an epistemic nature. The theory about your dog's behavior, for instance, undergoes processes of testing, proof, confirmation, support, and contradiction. In essence, there is no substantial evidence to suggest that there exists an interpretation of 'theory' that corresponds with the way conspiracy theories are commonly perceived.[11]

A third objection may pertain to the specificity of our examination of the term 'conspiracy theories'. The argument posits that individuals may use differing language when referring to conspiracy theories in the plural form as opposed to a singular conspiracy theory, or when discussing specific conspiracy theories such as the QAnon conspiracy theories or the flat earth theory. However, corpus data does not support this assertion. As presented in Table 7.6, the most commonly used verbs occurring before both phrases "VERB the conspiracy theory" and "VERB QAnon theories" belong to the category of spreading. It should be noted that this analysis does not suggest that all conspiracy theories are undeserving of the label 'theory'. For instance, theories surrounding the Watergate conspiracy may indeed be considered proper theories as characterized in this article. However, as Napolitano

[11] In a final attempt to address this issue, it might be proposed that there is a distinct sense of 'theory' unique to conspiracy theories. However, this perspective requires several critical justifications: (a) it must present a compelling and independent rationale for categorizing conspiracy theories as 'theories'; (b) it needs to clarify why specialists in the field of conspiracy theories have not recognized this unique interpretation, assuming it exists; and (c) it should convincingly argue against the necessity of re-engineering the term 'conspiracy theory' due to its idiosyncratic usage and implications.

and Reuter (2021) have proposed, such theories may no longer be classified as conspiracy theories in the contemporary understanding of the term.

Fourth, and finally, it may be argued that the connection between the language used to discuss conspiracy theories and the manner in which they are treated may not be entirely justified. A skeptic may posit that language use can often be misleading. While this is a valid point, the operationalization used in this analysis seems to accurately depict the actions taken with regard to theories, i.e., people discuss, apply and explain these theories. A more specific critique may therefore question whether the most frequently used verbs preceding 'conspiracy theories' accurately reflect the actions taken with regard to these theories. While it is unlikely that individuals would actively test, develop, and elaborate conspiracy theories but talk about them in a very different manner, here are two possible explanations why such a discrepancy might exist: (1) individuals may refrain from discussing the development and testing of conspiracy theories due to potential social repercussions, such as mobbing or backlash, and (2) those who approach conspiracy theories from a scientific perspective may be a minority and thus, their discourse is overshadowed by those who dismiss these theories. While these explanations would offer some interesting insight if they were true, they would also imply that the majority of individuals do not consider conspiracy theories to be theories. Are there any independent reasons to grant credibility to the minority's viewpoint? It seems that such a justification can only be established by granting the minority the status of experts. It would be difficult, to say the least, to argue that advocates of conspiracy theories deserve the label 'experts' in any substantial sense.

7.5 Re-Engineering *Conspiracy Theory* and Re-labelling 'Conspiracy Theory'

We started the paper with an exposition of the minimalist account which states that conspiracy theories are theories about conspiracies. Our corpus-linguistic investigations have uncovered that conspiracy theories bear minimal resemblance to established theories. Even when considering more lenient epistemic criteria for labeling claims as theories, akin to the case of fan theories, or "your theory", conspiracy theories fail to satisfy even this minimal requirement for classification as theories. Instead, they bear a close resemblance to misinformation and rumors.

What implications should be derived from the findings of our corpus studies? The discrepancy between the term 'conspiracy theory' and the results of our studies may immediately raise the call for some serious conceptual engineering. Conceptual engineering involves the process of refining individuals' concepts in order to improve their cognitive processes. The motivation behind this practice stems from the recognition that at times, it becomes necessary to enhance our conceptual frameworks in order to achieve various advantageous outcomes, whether they are theoretical, social, political, or of other nature (Isaac et al., 2022). One particularly well-known and extensively debated illustration is the redefinition of *planet* in 2006, wherein Pluto was excluded from the planetary category (for empirical data on the dissemination of re-engineered concepts like *planet*, see Landes and Reuter (2025). Notably, philosophers have taken up the task of refining specific concepts themselves (e.g., Haslanger, 2000; Napolitano and Reuter, 2021; Reuter & Brun, 2022; Scharp, 2013).

A natural question to ask at this point is what the theoretical, social, and political advantages are of such an engineering process in the case of *conspiracy theory*. On a theoretical level, the importance of precision is akin to the need for clarity in the classification of planets within the scientific community. Just as it is counterproductive to group dissimilar celestial bodies such as Pluto and Eris under the same category as Jupiter and Saturn, it is equally counterproductive to group conspiracy theories with proper theories. This delineation ensures the integrity of theoretical inquiry. From a social perspective, refining our terminology is imperative in order to dispel confusion and foster clearer communication. Given that 'conspiracy theory' is a term that permeates day-to-day conversation beyond academic circles, the likelihood of misunderstandings is considerably higher. By explicitly stating that claims regarding QAnon, Pizzagate, and Flat Earth are not theories in any relevant sense, we can mitigate the spread of false equivalencies. Politically, the atmosphere surrounding discussions on conspiracy theories is often charged and polarized. Some suggest that the term 'conspiracy theory' is employed to trivialize and dismiss contrary opinions (Shields, 2023). While reengineering or relabelling these concepts may not single-handedly transform these debates into rational discussions, the impact of such strategic communication adjustments on public dialogue should not be underestimated, as marketing strategies have often shown.

We need to be careful not to conflate the engineering of *terms* with the engineering of *concepts*, both of which have been argued to be part of the engineering process (see Landes, 2025, this volume, but see Isaac, 2020). Our results may only be taken to imply that conspiracy theories should not be defined as theories but rather as something else, like narratives or stories. In such a case, we would keep the label 'conspiracy theory' and thus only engage in the engineering of the concept *conspiracy theory*.

However, our results may also be interpreted to justify a more radical proposal, namely to change the very term that is currently used to refer to conspiracy theories. The current terminology, 'conspiracy theory', compositionally implies that they are indeed comprehensive and rigorous explanations. To avoid perpetuating the misconception that conspiracy theories are genuine theories, a modification of the label may be warranted. For instance, alternative labels such as 'conspiracy story' or 'conspiracy narrative' could be adopted. This latter proposal involves both conceptual and lexical engineering. Subsequently, we call these two options modest and radical proposal.

1. **Modest Proposal**: While we keep the label 'conspiracy theory', conspiracy theories should not be defined as theories or explanations but rather as stories or narratives that are based on speculative and unsupported ideas.

2. **Radical Proposal**: The term 'conspiracy theory' should be eliminated and replaced by the term 'conspiracy story' or 'conspiracy narrative'.

In the final section of this paper, we discuss both proposals and argue in favour of the radical proposal. More specifically, we will contend that 'conspiracy theory' should be replaced, and propose several suitable alternatives.

7.5.1 Re-Engineering Conspiracy Theory

The existing philosophical literature predominantly espouses the minimalist account, which explicitly defines conspiracy theories as theories. It seems advisable to also investigate how other authoritative sources approach this matter. Among the various online dictionaries commonly consulted, only Merriam-Webster defines conspiracy theory as a theory. Encyclopedia Britannica defines a conspiracy theory as an *attempt to explain* harmful events, Collins as a *belief* that a group of people are secretly trying to harm someone, and Cambridge Dictionary as a *belief* that an event or situation is the result of a secret plan. In comparison, when examining the definition of string theory, which serves as one of our control conditions in the corpus studies, all four dictionaries (Merriam-Webster, Encyclopedia Britannica, Collins, Cambridge Dictionary) consistently define it as a theory.

A cursory examination of these widely used online dictionaries highlights the lack of unanimity in defining conspiracy theory. This discrepancy suggests that reservations already exist regarding the characterization of conspiracy theories as theories. Consequently, it implies that conceptual engineering surrounding the term 'conspiracy theory' may already be underway.

Let us examine the notion of characterizing conspiracy theories as 'beliefs'. Both Collins and Cambridge Dictionary present a definition that aligns with this alternative perspective. However, we think that such a definition poses at least two significant problems. First, beliefs are typically understood as cognitive mental states. Theories, stories, views, and explanations, on the other hand, are not inherently cognitive entities themselves; instead, they constitute the content of our beliefs: We are convinced of theories, believe in stories, deliberate views, and think about explanations (see Uscinski, 2018, p.50). Thus, defining a conspiracy theory as beliefs, and not as the content of beliefs, fails to respect this distinction. Second, and of greater significance, we find that defining conspiracy theories as beliefs does not adequately reflect the empirical reality we have tried to depict in this article. Individuals frequently revise their beliefs in response to new evidence. Conversely, conspiracy theories are often constructed in a manner that renders counter-evidence inconsequential to their truth or falsity. This essential disparity undermines the proposition of classifying conspiracy theories as mere beliefs (but see Napolitano, 2021). Considering the weight of these arguments against defining conspiracy theories as beliefs, we are compelled to explore alternative approaches. Nonetheless, we acknowledge that such definitions still hold certain advantages over both the minimalist account and Merriam-Webster's definition.

Based on the findings from our corpus-analytical studies and the experimental studies conducted by Napolitano and Reuter (2021), we put forward the following re-engineered definition of conspiracy theory:

Definition: A conspiracy theory

(i) constitutes a story or narrative concerning a particular event or situation;
(ii) relies on speculative and unsubstantiated claims.[12]

Readers of this article may wonder, why we put forward an explicit definition of conspiracy theory given that we preferred a pragmatic approach in our corpus-linguistic analysis. The answer is rather straight-forward: We believe that definitions and characterizations can often be best investigated empirically by taking a practice-driven approach. Nonetheless, definitions and characterizations themselves are best formulated through an organized body of propositions. These propositions ideally constitute necessary and jointly sufficient conditions for the application of the target term. Such an ideal, however, often cannot be satisfied, given the vagueness, polyeidic and prototypical character of many concepts. In fact, we do not deny that conspiracy theories frequently include the suggestion of a conspiracy as an integral component of the event or situation. However, as the suggestion of an occurring conspiracy is not a necessary aspect of a conspiracy theory, our definition does not feature a proposition about a merely frequently occurring phenomenon.

We do owe the reader, however, a thorough justification for characterizing conspiracy theories as stories or narratives. In order to do so, we would like to delve into a related and, as we will see, instructive discourse on alternative labels of the German term 'Verschwörungstheorie'.

7.5.2 *The Re-labelling Discourse on 'Conspiracy Theory' in German*

The lack of discussion in the English-speaking realm regarding the ongoing usage of the term 'conspiracy theory' is quite surprising. However, the situation is markedly different when considering the discussions centered around the German term 'Verschwörungstheorie', which directly corresponds to 'conspiracy theory' and serves as the most prevalent term for denoting speculations regarding flat earth, lizard people, and other views in the German language. Over the past five years, alternative terms have emerged as viable alternatives (see especially Götz-Votteler and Hespers, 2019; Lamberty, 2020). These include 'Verschwörungshypothese' (conspiracy hypotheses), 'Verschwörungserzählung' (conspiracy narrative or con-

[12] Napolitano and Reuter also advocate for engineering *conspiracy theory* to include an epistemic evaluation. The second criterion thus nicely fits their approach. Yet, they do not focus on the theoretical aspect of conspiracy theories, which is something the first criterion of our definition deals with.

spiracy story), and 'Verschwörungsmythos' (conspiracy myth). Certain scholars (Götz-Votteler & Hespers, 2019; Lamberty, 2020, see also Reinalter, 2010) have proposed reserving the term 'Verschwörungshypothese' for serious and verifiable conspiracy accounts, while designating flat earth and QAnon conspiracy claims as 'Verschwörungserzählung'. Furthermore, Verschwörungsmythen constitute a subset of Verschwörungserzählungen, positing the existence of malevolent extraterrestrial beings, lizard people, and other mythical creatures.

Butter (2022) has defended the use of the term 'Verschwörungstheorie'. As his arguments are instructive and apply just as much to the question of whether we should keep the term 'conspiracy theory', let us respond to them one by one. First, Butter concurs with Hepfer's (2015) assertion that conspiracy theories share similarities with scientific theories. According to Hepfer, both conspiracy theories and scientific theories strive to acquire knowledge about the world. However, the findings of our two corpus analytic studies cast substantial doubt on the notion that conspiracy theories aim to generate knowledge. A crucial element of knowledge generation involves testing, developing, instructing, and engaging in discussions regarding their central claims. Our studies demonstrate that the majority of conspiracy theories fail to undergo such processes, resembling falsehoods and rumors instead. Consequently, there exists minimal convergence between conspiracy theories and (scientific) theories.

Second, Butter challenges the assertion made by critics that conspiracy theories lack falsifiability. According to Butter, it is evident that conspiracy theories can and often are refuted and proven false. While conspiracy theories are indeed falsifiable, Uscinski (2018, p. 49) correctly notes that "it is impossible to prove that a secret plot is not at work". The issue, therefore, lies not with conspiracy theories themselves, but rather with conspiracy theorists who refuse to acknowledge contradictory evidence and interpret it as further confirmation of the conspirators' deception. While we acknowledge the importance of distinguishing between conspiracy theories and conspiracy theorists, Butter's argument fails to consider the practical function and usage of conspiracy theories in comparison to scientific theories. Scientific theories are designed to provide plausible and evidence-based explanations for specific states of affairs, as demonstrated in our corpus analysis. People employ the term 'theory', and even more so, 'scientific theory', to fulfill that particular function. In contrast, individuals do not typically actively engage in developing, falsifying, refuting, or discussing conspiracy theories.[13] In other words, while we concur with Butter that conspiracy theories can be subject to falsification, most people neither approach conspiracy theories in a scientific manner nor do conspiracy theories serve the same purpose as scientific theories.

[13] There are, of course, exceptions to this, such as (ex-)Flat Earthers engaging in testing their theories. In the Netflix documentary "Behind the Curve", for example, Bob Knodel uses a 20,000$ gyroscrope in an attempt to falsify that the Earth rotates. However, such initiatives seem to be limited to a small fraction of conspiracy theorists.

Third, proponents of alternative labels such as 'Verschwörungserzählung' criticize the term 'conspiracy theory' for unjustifiably elevating conspiracy theories to the status of theories, despite their lack of deserving such recognition. Butter counters this argument by asserting that the term 'conspiracy theory' is deployed in a highly derogatory manner, targeting those who espouse particular beliefs. We believe that both claims are not necessarily contradictory. While it is true that 'conspiracy theory' carries a negative connotation (as highlighted by Napolitano and Reuter, 2021), it creates the impression of being a theory, albeit a flawed one. It would be prudent for us to refrain from even considering conspiracy theories as theories. Ultimately, reengineering the concept conspirary theory can direct people's communicative strategies away from the disparaging use of 'conspiracy theory', and thus further our understanding of why proponents of conspiracy theories hold such irrational viewpoints.

7.5.3 Which Term Best Replaces 'Conspiracy Theory'

We are now in a position to make our case for the more radical proposal, which goes beyond merely re-engineering the concept of *conspiracy theory* and extends to relabel those viewpoints that have traditionally been referred to as 'conspiracy theories'. As readers may anticipate, our proposed alternative labels are directly influenced by the German term 'Verschwörungserzählung'. Although 'Verschwörungstheorie' remains the more commonly used expression, the successful adoption of this term within German discourse serves as tangible evidence that relabeling is not only a theoretical possibility but also a practical undertaking that can be effectively implemented.

Thus, we would like to propose to replace the term 'conspiracy theory' with the more accurate and theoretically sound terms 'conspiracy story' or 'conspiracy narrative'.[14] In fact, there are emerging approaches in this direction in the English literature as well. One of the most promising approaches treats conspiracy theories as fictional narratives rather than genuine beliefs: Non-doxastic fiction accounts of conspiracy theory (e.g., Heering, 2022; Ichino and Juha Räikkä, 2021; Ichino, 2022; Römer, 2021, see also Fenster, 2008) emphasize the role of narrative construction in conspiracy theories, highlighting their function as storytelling devices that shape collective understanding and interpretations of complex events. These accounts shed light on the psychological and social functions that conspiracy theories serve, providing individuals and groups with explanations, meaning, identity, and empowerment, even without a commitment to their literal truth. They acknowledge

[14] Why do we propose two terms instead of settling on a single recommendation? The straightforward answer is that we consider 'story' to be the more fitting term, but we acknowledge the potential negative ramifications. 'News stories' generally encompass earnest portrayals of real-world events. There exists a concern that proponents of outlandish conspiracy theories could exploit the term 'conspiracy story' to undermine the credibility of legitimate news articles.

the epistemic limitations of conspiracy theories, recognizing their lack of empirical evidence, logical coherence, and scientific methodology. By considering these aspects, non-doxastic accounts avoid treating conspiracy theories as equivalent to established knowledge claims and provide a more nuanced understanding of their nature and impact.

Ichino (2022) and Heering (2022) have put forth a narrative model, suggesting that when conspiracy theorists *believe* in a conspiracy, they are engaged in a complex game of Waltonian make-believe (Walton, 1990). They imagine a world where the conspiracy is true, while the content of their alleged beliefs serve as explanatory elements within this fictional narrative. The conspiracy itself serves as an explanation *within the story*, rather than an attempt to explain a real event through fiction. One of the major differences between belief and make-believe is that "[b]elief responds to evidence and reasons in ways in which make-believe doesn't, and this means[...] that belief undergoes constraints of coherent inferential integration that do not equally apply to make-believe." (Ichino, 2022, p. 246) According to Walton (1990), make-believe involves engaging in a form of pretense or imaginative play, where individuals temporarily suspend their beliefs and adopt a fictional perspective. In this state of make-believe, people willingly enter into an alternative reality, accepting certain premises or assumptions that are not considered true in the real world. This framework can elucidate how an individual can actively participate as a fully functioning and rational member of society while, at the same time, immersing themselves deeply in the realm of conspiracy theories—reserving their engagement in make-believe for salient contexts. In instances where conspiracy theories exert substantial influence over a person's life, it is plausible that reality has been significantly supplanted.

While the narrative account of conspiracy theory seems promising, let us briefly touch on two possible objections. First, one may argue that a fictional account of conspiracy theories fails to explain their subversive and heterodox potential in challenging official views. Heering (2022) argues that conspiracy theories do not necessarily require an equivalent counter-hypothesis to challenge orthodox explanations. What is important for heterodoxy is that the topoi of conspiracy narratives—the abuse of power, the undemocratic influence of interest groups—are heard in society. Narratives are well-suited for disseminating the rebellious attitude and values needed for cultivating heterodoxy, often more so than actual hypotheses about specific conspiracies. Hence, the allure and subversive potential of conspiracy theories is preserved and emphasized in the narrative model.

Second, one might object that by embracing a non-doxastic account of conspiracy theories, calling them out as false, unfounded, or in any other way epistemically irrational, is a category mistake. However, according to Ichino, this is not inherently problematic: "to say that the attitudes towards conspiracy theories are not epistemically irrational does not amount to saying that there is nothing whatsoever wrong with them and that those who endorse them are immune from criticism of any sort." (Ichino, 2022, p. 256) Within a narrative framework of conspiracy theories, critique can still be directed at moral, prudential, or political aspects. However, this answer will not satisfy those who doubt that such a category error exists. It could well be argued that we are giving away too much if we can no longer treat conspiracy theories as false judgments. However, perhaps this problem can be circumvented metalinguistically. For it seems unproblematic to say that the statement "the floor is lava" is false outside of the eponymous game (i.e. the collective make-believe). Analogously, it seems felicitous to say that someone's make-believe involving a conspiracy theory is false *if* they behave like it is true outside of the fictional setting.[15]

Particularists like Dentith (2019), Pigden (2007) and many others have tiredlessly argued that many conspiracy theories, such as the Watergate scandal or the default 09/11 conspiracy account are true and based on sound evidence. Consequently, according to their perspective, these theories should not be dismissed as mere 'conspiracy stories' or 'conspiracy narratives'. We concur with this viewpoint. Nonetheless, as highlighted by Napolitano and Reuter (2021), it is more appropriate to categorize such theories (and we wholeheartedly embrace the term 'theories') as 'conspiratorial explanations': "[t]he descriptive *conspiratorial explanation*—rather than a descriptive engineering of conspiracy theory—could be employed for those philosophical projects which necessitate a neutral definition, such as comparing explanations that involve conspiracies to other types of explanations, which have typically been the object of investigation [...]" (2021, p. 2058)

Before drawing to a close, let us revisit the proposal we have put forth. The arguments articulated in this section strongly advocate for the adoption of the terms 'conspiracy story' and 'conspiracy narrative' as substitutes for the inadequate expression 'conspiracy theory'. It is important to note that these same arguments also support the more moderate proposal of defining conspiracy theories as stories or narratives pertaining to a particular event. For those who may not be inclined to embrace the radical proposition, we suggest that a more precise definition of conspiracy theory should not refer to 'theories' or 'explanations' but rather 'stories' or 'narratives'.

7.6 Conclusion

In this paper, we conducted a set of corpus analyses of the composite term 'conspiracy theory'. The results of our studies show that conspiracy theories are not discussed and tested like genuine theories but rather promoted and spread like falsehoods and rumours. We then proposed a relabeling of the predominant and most significant subset of conspiracy theories as 'conspiracy stories' or 'conspiracy narratives'. This proposal finds support not only in the empirical evidence expounded in this paper but also derives motivation from the (partially) accomplished adoption of the German counterpart term 'Verschwörungserzählung', as well as the theoretical benefits associated with a non-doxastic perspective on conspiracy theories.

[15] We would like to thank an anonymous reviewer for pointing this out.

References

Abend, G. (2008). The meaning of 'theory'. *Sociological theory, 26*(2), 173–199.

Alfano, M. (2024). Semantic network analysis for philosophers. In S. Kornmesser & A. M. Bauer et al. (Eds.), *Experimental philosophy for beginners*. Springer.

Basham, L., & Dentith, M.R.X. (2016). Social science's conspiracy theory panic: Now they want to cure everyone. *Social Epistemology Review and Reply Collective, 5*(10), 12–19.

Baumgartner, L. (2022). Why are reckless socks not (more of) a thing? Towards an empirical classification of evaluative concepts. *Proceedings of the Annual Meeting of the Cognitive Science Society, 44*(44).

Baumgartner, J., Zannettou, S., Keegan, B., Squire, M., & Blackburn, J. (2020). The pushshift Reddit dataset. arXiv:2001.08435.

Butter, M. (2022). Verschwörungstheorien: Eine Einführung. In K. Müller & C. Kirchberg (Eds.), *Verschwörungstheorien* (pp. 12–27). Gegen Vergessen - Für Demokratie e.V.

Carnap, R. (1966). *Philosophical foundations of science.* Basic Books.

Cassam, Q. (2019). *Conspiracy theories,* Polity Press.

Cinelli, M., Etta, G., Avalle, M., Quattrociocchi, A., Marco, N. D., Valensise, C., Galeazzi, A., & Quattrociocchi, W. (2022). Conspiracy theories and social media platforms. *Current Opinion in Psychology, 47*, 101407.

Coady, D. (2003). Conspiracy theories and official stories. *International Journal of Applied Philosophy, 17*(2), 197–209.

Cohnitz, D. (2018). On the rationality of conspiracy theories. *Croatian Journal of Philosophy, 18*(2), 351–365.

Davies, M. (2016). Corpus of News on the Web (NOW). Available online at https://www.english-corpora.org/now/.

Dentith, M. R. X. (2019). Conspiracy theories on the basis of the evidence. *Synthese, 196*(6).

Dentith, M. R. X. (2022). Suspicious conspiracy theories. *Synthese, 200*(3), 243.

Duetz, J. C. M. (2022). Conspiracy theories are not beliefs. *Erkenntnis*, 1–15.

Duetz, J. C. M. (2023). What does it mean for a conspiracy theory to be a 'theory'? *Social Epistemology*, 1–16.

Đorić, M. (2020). Sociopolitical crises in the times of Covid-19. *Security Crises in the 21st Century and how to manage them social and security aspects* (Vol. 140).

Engel, K., Hua, Y., Zeng, T., & Naaman, M. (2022). Characterizing Reddit participation of users who engage in the qanon conspiracy theories. *Proceedings of the ACM on Human-Computer Interaction*, 6, 1–22.

Feldman, S. (2011). Counterfact conspiracy theories. *International Journal of Applied Philosophy, 21*(1), 15–24.

Fenster, M. (2008). *Conspiracy Theories, Secrecy and Power in American Culture.* Revised Edition. Minneapolis, MN.

Frindte, W., & Frindte, I. (2023). "But Otherwise But Otherwise. All Lies!" (Rio Reiser, 1986)—From the charm of conspiracy theories. In *Support in Times of No Support: A Social Psychological Search for Traces* (pp. 157–164). Springer Fachmedien Wiesbaden.

Götz-Votteler, K., & Hespers, S. (2019). *Alternative Wirklichkeiten? Wie Fake News und Verschwörungstheorien funktionieren und warum sie Aktualität haben.* Bielefeld.

Harris, K. (2018). What's epistemically wrong with conspiracy theorising? *Royal Institute of Philosophy Supplement, 84*, 235–257.

Haslanger, S. (2000). Gender and race: (What) are they?(What) do we want them to be?. *Noûs, 34*(1), 31–55.

Heering, D. (2022). Alethische und Narrative Modelle von Verschwörungstheorien. *Zeitschrift für Praktische Philosophie, 9*(2), 143–174.

Hempel, C. (1966). *Philosophy of natural science.* Prentice-Hall.

Hepfer, K. (2015). *Verschwörungstheorien. Eine philosophische Kritik der Unvernunft.* Bielefeld.

Huneman, P., & Vorms, M. (2018). Is a unified account of conspiracy theories possible? *Argumenta Oeconomica Cracoviensia, 3*, 49–72.
Ichino, A. (2022). Conspiracy theories as Walt-Fiction. In P. Engisch, & J. Langkau (Eds.), *The Philosophy of Fiction* (pp. 240–261). Taylor & Francis.
Ichino, A., & Räikkä J. (2021). Non-doxastic conspiracy theories. *Argumenta, 7*(1), 247–263.
Isaac, M. G. (2020). How to conceptually engineer conceptual engineering?. *Inquiry*, 1–24.
Isaac, M. G., Koch, S., & Nefdt, R. (2022). Conceptual engineering: A road map to practice. *Philosophy Compass, 17*(10), e12879.
Keeley, B. (1999). Of conspiracy theories. *Journal of Philosophy, 96*, 109–126.
Lamberty, P. (2020). Zwischen Theorien und Mythen: eine kurze begriffliche Einordnung. *Informationen zur politischen Bildung*, 344.
Landes, E. (2025, this volume). How language teaches and misleads: "Coronavirus" and "Social Distancing" as case studies. In M. G. Isaac, K. Scharp, & S. Koch (Eds.), *New Perspectives on Conceptual Engineering*.
Landes, E., & Reuter, K. (2025). Conceptual revision in action. *Review of Philosophy and Psychology*, 1–30.
Meylan, A., & Reuter, K. (2025). On being ignorant. *Canadian Journal of Philosophy*, 1–18.
National Academy of Sciences. (1998). *Teaching about evolution and the nature of science*. Joseph Henry Press.
Napolitano, M. G. (2021). Conspiracy theories and evidential self-insulation. *The Epistemology of Fake News*, 82–106.
Napolitano, M. G., & Reuter, K. (2021). What is a conspiracy theory? *Erkenntnis*, 1–28.
Napolitano, M. G., & Reuter, K. (2023). Is *conspiracy theory* a case of conceptual domination? *Social Epistemology Review and Reply Collective, 12*(11), 74–82.
Pigden, C. (2007). Conspiracy theories and the conventional wisdom. *Episteme, 4*(2), 219–232.
Popper, K. (1963/2014). *Conjectures and refutations: The growth of scientific knowledge*. Reprinted by Routledge, 2014.
Räikkä, J. (2018). Conspiracies and conspiracy theories: An introduction. *Argumenta, 3*(2), 205–216.
Reinalter, H. (2010). *Die Weltverschwörer: Was Sie eigentlich alles nie erfahren sollten*. Ecowin Verlag.
Reuter, K. (2011). Distinguishing the appearance from the reality of pain. *Journal of Consciousness Studies, 18*(9–10), 94–109.
Reuter, K., & Brun, G. (2022). Empirical studies on truth and the project of re-engineering truth. *Pacific Philosophical Quarterly, 103*(3), 493–517.
Reuter, K., & Baumgartner, L. (2024). Corpus analysis: A case study on the use of 'conspiracy theory'. In S. Kornmesser, & A. M. Bauer et al. (Eds.) *Experimental philosophy for beginners*. Springer.
Reuter, K., Baumgartner, L., & Willemsen, P. (2023). Tracing thick and thin concepts through corpora. *Language and Cognition, 16*(2), 263–282.
Römer, D. (2021). Verschwörungstheorien als argumentative Narrative. *Zeitschrift für Literaturwissenschaft und Linguistik, 51*, 281–298.
Rosch, E. (2024). Principles of categorization. In *Cognition and categorization* (pp. 27–48). Routledge.
Scharp, K. (2013). *Replacing truth*. Oxford University Press.
Shahsavari, S., Holur, P., Wang, T., Tangherlini, T. R., & Roychowdhury, V. (2020). Conspiracy in the time of corona: automatic detection of emerging covid-19 conspiracy theories in social media and the news. *Journal of Computational Social Science, 3*, 279–317.
Shields, M. (2023). Conceptual engineering, conceptual domination, and the case of conspiracy theories. *Social Epistemology*, 1–17.
Suppes, P. (1960). A comparison of the meaning and uses of models in mathematics and the empirical sciences. *Synthese, 12*, 287–301.

Sytsma, J., Bluhm, R., Willemsen, P., Reuter, K. (2019). Causal attributions and corpus analysis. In E. Fischer, & M. Curtis (Eds.), *Methodological advances in experimental philosophy* (pp. 209–238).

Tsapos, M. (2023). Who is a conspiracy theorist?. *Social Epistemology*, 1–10.

Uscinski, J. (2018). *Conspiracy theories and the people who believe them*. Oxford University Press.

Walton, K. (1990). *Mimesis as Make-Believe: On the Foundations of the representational arts*. Harvard University Press.

Willemsen, P., Baumgartner, L., Frohofer, S., & Reuter, K. (2021). Examining evaluativity in legal discourse: A comparative corpus-linguistic study of thick concepts. In S. Magen, & K. Prochownik (Eds.), *Advances in experimental philosophy of law*. Bloomsbury Publishing.

Winther, R. (2020). The structure of scientific theories. In E. Zalta (Ed.) *The Stanford Encyclopedia of Philosophy*.

Wittgenstein, L. (1953/2010). *Philosophical investigations*. John Wiley & Sons.

van Fraassen, B. (1989). *Laws and symmetry*. Oxford University Press.